Smart Card
Developer's Kit

Scott B. Guthery

Timothy M. Jurgensen

MACMILLAN
TECHNICAL
PUBLISHING
U·S·A

Macmillian Technical Publishing, Indianapolis, Indiana

SMART CARD DEVELOPER'S KIT

By Scott B. Guthery and Timothy M. Jurgensen

Published by:
Macmillan Technical Publishing
201 West 103rd Street
Indianapolis, IN 46290 USA

Copyright © 1998 by Macmillan Technical Publishing

Printed in the United States of America
1 2 3 4 5 6 7 8 9 0

International Standard Book Number:
1-57870-027-2

Library of Congress Number: 97-81197

Warning and Disclaimer

Publisher	*Don Fowley*
Associate Publisher	*Jim LeValley*
Executive Editor	*Ann Trump Daniel*
Brand Manager	*Alan Bower*
Director of Editorial Services	*Carla Hall*
Managing Editor	*Caroline Roop*

Development Editor
Kitty Wilson Jarrett

Project Editors
Sherri Fugit
Brad Herriman

Copy Editor
Susan Hobbs

Software Specialist
Jack Belbot

Team Coordinator
Amy Lewis

Manufacturing Coordinator
Brook Farling

Book Designer
Anne Jones

Cover Designer
Aren Howell

Production Team Supervisor
Andrew Stone

Graphics Image Specialists
Steve Adams
Debi Bolhuis
Kevin Cliburn
Sadie Crawford
Wil Cruz
Tammy Graham
Oliver Jackson

Production Analysts
Dan Harris
Erich J. Richter

Production Team
Jeanne Clark
Christy M. Lemasters
Julie Searls
Sossity Smith
Heather Stephenson

Indexer
Sandy Henselmeier

About the Authors

Scott B. Guthery leads the cryptographic device project at CertCo. He holds a Ph.D. in probability and statistics from Michigan State University and worked for Bell Laboratories and Schlumberger before joining CertCo. He holds two patents for his work on Schlumberger's well logging system and led the team that invented and developed the Java Card smart card. He has published articles in number theory, programming languages, real-time systems, and network protocols.

Timothy M. Jurgensen is a senior member of the technical staff at Schlumberger's Austin Products Center. He holds a Ph.D. in nuclear physics from Rice University. He has been with Schlumberger for more than 25 years, involved in development activities for wide area computer networks deployed to support Schlumberger's Oilfield Services business segment. He has been particularly active in the development of small aperture (VSAT) satellite-based communication systems and in the provision of high-security computer networks. He has published articles regarding wide area computer networks, satellite-based communication systems, data management, and network security infrastructures.

Trademark Acknowledgments

All terms mentioned in this book that are known to be trademarks or service marks have been appropriately capitalized. Macmillan Technical Publishing cannot attest to the accuracy of this information. Use of a term in this book should not be regarded as affecting the validity of any trademark or service mark.

Solo, Cyberflex, Multiflex, and Payflex are trademarks of Schlumberger. Java and Java Card are trademarks of Sun Microsystems. Windows, Windows 95, and Windows NT are trademarks of Microsoft Corporation. EZ Formatter, EZ Formatter Trial, and EZ Component are trademarks of Strategic Analysis, Inc. SnareWorks is a trademark of Intellisoft Corp. Visa Cash is a trademark of Visa. AAdvantage is a trademark of American Airlines.

Dedications

Scott B. Guthery—to M.J.K.

Timothy M. Jurgensen—to Becky, Sarah, and Miriam.

Acknowledgments

The authors would like to thank Maria Nekam for her help with the design of the book's cover and smart card. Ms. Nekam brought a healthy dose of ingenuity and enthusiasm to the various graphic designs related to the book.

The authors would also like to thank Henry Tumblin, who did an outstanding technical review on the book. Henry was a faithful and demanding representative of you, the reader. The book benefited greatly from his attentions.

The authors would like to acknowledge the generous support of Schlumberger during the development of this book. The members of the Austin Smart Card Team have been extremely helpful with their suggestions, review, and general support. Special thanks go to Danita Moseles, Ksheerabdhi Krishna, Neville Pattinson, Bertrand du Castel, Ed Dolph, Marc Valderrama, Pam Saegert, and Mike Montgomery for their active support. The encouragement and support of the entire Smart Cards and Systems group at Schlumberger is most appreciated as well.

Any errors that remain—particularly the ones that may seem to jump off the page at you when the book is opened at random—are entirely the responsibility of the authors. Your efforts to call these to our attention via email will be greatly appreciated. Any opinions expressed in the book are those of one or both of the authors and do not necessarily reflect the opinions of Schlumberger.

Contents at a Glance

Table of Contents

Part II Smart Card Software Development

Part III Smart Card Application Examples

Part IV   Appendixes

FOREWORD

Five years ago the annual issue of smart cards was measured in the tens of millions—described by the cynical as "a solution looking for a problem." Today the annual issue of smart cards is measured in the hundreds of millions, and by the year 2002 smart card issue will be measured in billions of cards.

Smart cards are a fundamental and increasingly important part of any delivery strategy, enabling existing and new products and services to be delivered more efficiently, more conveniently, and more securely.

Smart cards have been adopted by governments to counter benefit fraud, by payment associations such as MasterCard and Visa (via their members) to secure payment whether in-person or over the telephone or the Internet, by mobile telephone companies to secure access to their services and to make those services more useful and user friendly, by retailers and airlines for loyalty programs, by network managers to secure access, and by companies such as Mondex and Proton to enable the new payment product—electronic cash.

Smart cards are the hottest tool in the IT armory, with a unique and complementary set of properties: portability, highly secure and tamper-resistant data storage capability (for cryptographic keys and other data), programmable, affordable, and well-accepted by consumers. This is clearly evident as we see the smart card business moving from being a separate industry to being an extension of the IT industry and with companies such as Microsoft and Sun Microsystems becoming increasingly engaged.

Historically, smart cards were perhaps only economical and practicable for large-scale projects with thousands of cards. However, this is also changing; it is now entirely viable to use a smart card for projects requiring a few dozen cards—such as a company security card or the Internet access token for a club or group. This change has been enabled through the development of operating systems with open APIs such as MULTOS and Java Card.

This book provides anybody who is interested in using smart cards with a ready entry point, speeding the learning process and lowering—if not eliminating—many of the traditional barriers to entry.

Smart Card Developer's Kit steers the reader through the constraints as well as the benefits of the smart card, navigating through the myriad of specifications and standards, evaluating the development tools, discussing the security issues and implications, and providing some useful, illustrative examples. The many references to contacts, suppliers, and sources of information will save much time and frustration.

The book does not solely focus on on-card elements, but also explains the role and importance of the reader-side part of the solution. It discusses many of the key standards and specifications, including, for example, the PC-to-smart-card (PC/SC) interface.

This book combines an eminently helpful combination of theory, analysis, and information supported by practical examples, suggestions, and advise—no doubt derived from the hard-won experience of the authors. The inclusion of the CD and Multiplex card will allow readers to consolidate their new-found knowledge by allowing them to develop and test actual smart card applications.

Although the book is perhaps aimed at the IT community and programmers (it forms an excellent conversion text for those switching from general IT into smart cards), it will certainly also appeal to product and project managers as a point of education and reference.

I wish that *Smart Card Developer's Kit* had been available when I first joined the industry!

Nick Habgood
Chief Executive, MAOSCO Ltd.
email: nick.habgood@multos.com
Internet: www.multos.com

PREFACE

The smart card as a general-purpose software application platform has been shrouded in secrecy for almost the entire 25 years of its existence. Early deployment of smart card–based systems with their requirements for large infrastructure investment was facilitated by large, often governmental card issuers. Applications were concentrated in the transportation, telecommunications, and financial business sectors. Security concerns among card issuers in all these areas led to significantly restricted flows of information to the general computing and networking communities regarding smart card systems, their software architectures and development environments, along with their hardware technology. Coupled with an overarching concern for low cost and high reliability which tended to be best served by "mature" technologies, this has caused the evolution of the smart card, as a general-purpose computing platform, to fall far behind the pace of advancement found in other arenas such as desktop computing.

Just as the public Internet and the World Wide Web have opened up the discussion of encryption technology, so have they brought the capabilities of the smart card to the attention of the open market and technical and product innovators. This book reaches beyond generic discussions of smart card capabilities and programs and gets down to the bits, bytes, and details of real-life smart card programming. The goal of the book is to put into the reader's hands a significant body of the information and resources needed to build and field new and innovative smart card applications.

The Smart Card Computer

It is a common perception that the computer in a smart card is so small and so insignificant as to be useful for doing little more than adding and subtracting small integers and comparing byte-size values. From the point-of-view of people dealing with today's 100 MIP, 100 MB desktop machines, this perception is certainly understandable.

The authors of this book came to smart cards after spending 17 and 23 years, respectively, involved with the building, deployment, operations, and support of truck-mounted oil well logging systems using computers with roughly the same computing power of a smart card. Coming from an environment in which one could log an oil well on the North Slope of Alaska with a quarter MIP, it seemed plausible that one could certainly do some very interesting things with a quarter MIP in a nice warm shopping mall, a corporate office, or a student's home PC.

With a self-assurance heavily rooted in a profound ignorance of the smart card industry, but enticed by the seemingly glaring opportunities open to smart card applications, we looked for the signposts which would guide us into this new (at least for us) frontier. Finding them typically sparse, usually covered by myriad non-disclosure agreements, sometimes behind locked doors, and often written in German and French, we decided to take a stab at providing something of a nuts-and-bolts guidebook that we more often found in other disciplines of the computing world.

As firm believers in the conventional wisdom that, historically, killer applications for a technology do not come from the providers of the technology, the authors set about to contribute their efforts to opening up smart card software development opportunities to software communities outside the normal smart card world. Scott Guthery lead the small team at Schlumberger's Austin System Center that invented the Java Card smart card and Tim Jurgensen participated in the PC/SC Workgroup's efforts that are encouraging an open specification–based smart card infrastructure in personal computer operating systems.

It is our fervent hope that you will find the information contained herein to be a substantial aid in understanding smart card technology as it can apply to the general computing and networking environments and perhaps even sow the seeds of the killer app that's lurking out there somewhere.

Organization of the Book

The book is divided into three major parts: Smart Card Background and Basics, Smart Card Software Development, and Smart Card Application Examples.

Part I, "Smart Card Background and Basics," consists of five chapters that describe the smart card computer from a programmer's perspective. Chapter 1 is a general overview of smart card software with particular emphasis on what makes smart card software different from other kinds of software. Chapter 2 discusses the general physical properties of a smart card so that the smart card programmer can understand why the hardware resources of a smart card are the way they are and how they were intended by their designers to be used. Interoperability of smart cards is a constant concern, although perhaps an underachieved reality of the smart card industry. Chapter 3 familiarizes the reader with the major standards and specifications that are the foundation sourcebooks of smart card programmers. Chapters 4 and 5 take a detailed look at the commands that you send to a smart card to make it do what you want. Chapter 4 covers communication with industry standard commands, while Chapter 5 zeros in on the particular properties of the Schlumberger 3K Multiflex card included with the book.

Part II, "Smart Card Software Development," includes four chapters that catalog the many software tools that are available to support smart card software development projects. Chapter 6 introduces the various software development and debugging aids that a smart card programmer might use in the course of a software project. The chapter includes names, telephone numbers, email addresses and URLs of the tool providers. Chapter 7 is about application programming interfaces available to host computer applications that want to incorporate the capabilities of smart cards. The chapter includes thumbnail sketches of the many APIs together with a description of strengths, weaknesses, and current developments. Chapter 8 is about writing software that runs on the smart card itself. While writing software for the card is definitely the exception rather than the rule right now, we expect to see more card software written as application writers discover the capabilities of the smart card. Finally, Chapter 9 focuses on smart cards as they impinge on the security goals that are the focus of so much Internet and intranet development activity.

Part III, "Smart Card Application Examples," walks you through the development of two smart card applications that are representative of two application areas that are particularly popular today: loyalty programs in Chapter 10 and electronic commerce in Chapter 11. The 3K Multiflex card included with the book is used to develop these applications and thus the example applications serve as further instruction on how to harness the Multiflex card.

Part IV, "Appendixes," is a useful reference to smart card commands. Appendix A lists the ISO 7816-4 command set, and Appendix B lists the Multiflex command set.

The Schlumberger 3K Multiflex Card

The Schlumberger 3K Multiflex smart card included in this book is a general-purpose off-the-shelf smart card that supports typical industry-standard smart card commands together with some additional commands that are particularly useful for electronic purses and loyalty programs. The book includes a full chapter and an appendix documenting the card together with an extended loyalty program application that illustrates the use of the card.

Corrections, Updates, and Additions

Despite the efforts of many, there are undoubtedly errors in the book. Furthermore, we know first-hand that the information we have sought to include in the books is subject to change almost as quickly as we record it. Finally, our time resources were finite so we have surely overlooked material that should have been included.

We encourage you to let us know via email about shortcomings you find in the book. The authors will be maintaining a Web page at http://www.scdk.com/ to distribute corrections, updates, and additions to the information presented herein.

Scott B. Guthery
Boston, Massachusetts
sguthery@tiac.net

Timothy M. Jurgensen
Austin, Texas
tjurgensen@austin.asc.slb.com

December 1997

PART I

SMART CARD BACKGROUND AND BASICS

CHAPTER 1

SMART CARD PROGRAMMING

A smart card is a portable, tamper-resistant computer with a programmable data store. It is the exact shape and size of a credit card, holds 16 KB or more of sensitive information, and does a modest amount of data processing as well. The central processing unit in a smart card is typically an 8-bit microcontroller that has the computing power of the original IBM PC. To make a computer and a smart card communicate, you place the card in or near a smart card reader, which is connected to the computer.

After a period of some 20 years of arrested development, smart cards are beginning to evolve. Memory sizes are increasing and processor architectures are moving to 16-bit and 32-bit configurations. This book is about software development for today's widely available 8-bit microcontroller smart cards.

A Brief History

The smart card— a term coined by French publicist Roy Bright in 1980—was invented in 1967 and 1968 by two German engineers, Jürgen Dethloff and Helmut Gröttrupp. Dethloff and Gröttrupp filed for a German patent on their invention in February 1969 and were finally granted patent DE 19 45 777 C3, titled "Identifikanden/Identifikationsschalter," in 1982. Independently, Kunitaka Arimura of the Arimura Technology Institute in Japan filed for a smart card patent in Japan in March 1970. The next year, in May 1971, Paul Castrucci of IBM filed an American patent titled simply "Information Card" and on November 7, 1972, was issued U.S. Patent 3,702,464. Between 1974 and 1979 Roland Moréno, a French journalist, filed 47 smart card–related patent applications in 11 countries and founded the French company Innovatron to license this legal tour de force. Two brief but excellent histories of the early days of the smart card by Klaus H. Knapp of the Eduard Rhein Foundation and Robin Townend of MasterCard International can be found at

http://www.eduard-rhein-foundation.de/html/t96.html

and

http://www.smartcard.co.uk/finance1.html.

Smart cards cost between $1 and $20, depending primarily on the size of the memory in the card and the software functionality included. Smart card software usually includes a rudimentary on-board operating system with file system, communication, authorization, encryption, and access control primitives. Smart cards are particularly useful components of computer systems that need to address data security, personal privacy, and user mobility requirements.

Smart card programming is characterized by a constant and overarching concern for two system requirements: data security and data integrity. Data security means that a data value or a computational capability contained on the card can be accessed by those entities that are authorized to access it and not accessed by those entities that are not authorized to access it. Data integrity means that at all times the value of information stored on a card is defined; the value is not corrupted, even if power to the smart card is cut during a computation involving some piece of information stored on the card.

Unlike many software applications commercial programmers deal with daily, smart card applications are typically public systems. This means first that smart cards are used in settings and situations in which using a computer is not the first thing on the user's mind. Furthermore, the smart card computer must fit seamlessly and, to the greatest extent possible, unnoticed into existing behaviors and relationships. Paying for a newspaper with electronic money on a smart card should, for example, be very much like paying for the newspaper with cash.

Furthermore, unlike applications that are run on corporate computers in physically guarded environments and on private networks, smart card computers are "on the street" and subject to attack by a range of interests and agendas that are not fully known, let alone understood, by the smart card programmer and system designer.

The amount of data processed by a smart card program is usually quite small and the computations performed are typically quite modest. Subtracting 50 cents from a smart card's electronic purse, for example, entails neither many numbers nor much arithmetic. However, making sure that the expectations of all the parties to this transaction—the cardholder, the card issuer, and the merchant—are met during and after the transaction places an unfamiliar and relatively unique set of demands on software system designers and computer programmers. The merchant expects to be exactly 50 cents richer, the cardholder expects to be exactly 50 cents poorer, and the card issuer expects that the smart card will be left in a consistent state and not have leaked information about its inner workings.

Smart Versus Memory Versus Logic Cards

Two other credit card–sized devices contain integrated circuits; because of their identical size and similar uses, they are sometimes confused with smart cards. These are memory cards and logic cards. Memory cards contain only memory and logic cards are memory cards with some added circuitry to provide some data security functions. European telephone cards, transportation fare cards, and some medical record cards are examples of the uses of these cards. Since neither memory nor logic cards contain a general-purpose programmable processor for which software can be written, they are not explicitly considered in this book. Nevertheless, many of the discussions about systems that use smart cards apply equally well to memory and logic cards. Memory cards and logic cards should certainly be considered by system designers in building systems that include a portable personal data store.

Smart Card Software

There are fundamentally two types of smart card software, some examples of which are listed in Table 1.1:

- *Host software*, which is software that runs on a computer connected to a smart card. Host software is also referred to as *reader-side software*.

- *Card software*, which is software that runs on the smart card itself. As a counterpart of reader-side software, card software is also referred to as *card-side software*.

Host Software

Most smart card software is host software. It is written for personal computers and workstation servers, accesses existing smart cards and incorporates these cards into larger systems. Host software will typically include end-user application software, system-level software that supports the attachment of smart card readers to the host platform, and system-level software that supports the use of the specific smart cards needed to support the end-user application. In addition, host software includes application and utility software necessary to support the administration of the smart card infrastructure.

Host software is usually written in one of the high-level programming languages found on personal computers and workstations—C, C++, Java, BASIC, COBOL, Pascal, or FORTRAN—and linked with commercially available libraries and device drivers to access smart card readers and smart cards inserted into them. In constrast, card software is usually written in a safe computing language such as Java, machine-level language such as Forth, or assembly language.

Card Software

Card software is the software that runs on the smart card itself. It is usually classified as operating system, utility, and application software, much as is the case with host software. For many applications, rather generic smart cards with their general on-card software will suffice; special software for the card is not required. Where application-specific card software is required, it is typically written either in assembly language for the chip architecture of the microprocessor found embedded in the smart card or in a higher-level language that can be interpreted directly on the card or compiled into card assembly language and loaded onto the card.

It is useful to occasionally further categorize smart card software into application software or system software. Application software uses the computational and data storage capabilities of a smart card as if they were those of any other computer and is relatively unaware of the data integrity and data security properties of the smart card. These are of more concern to the person using the card than to the application software accessing it. System software, on the other hand, explicitly uses and may contribute to and enhance the data integrity and data security properties of the smart card.

Host application software substitutes the smart card for an alternative implementation of the same functionality (for example, when an encryption key or a medical record is kept on a smart card rather than on a hard disk file on the local computer or in a central database on a server). Host system software harnesses the unique and intrinsic computing and data storage capabilities of the smart card by sending data and commands to it and by retrieving data and results from it.

Card application software is typically used to customize an existing off-the-shelf smart card for a particular application (see Table1.1) and amounts to moving some functionality from host application software onto the card itself. This may be done in the interest of efficiency—in order to speed up the interaction between the host and the card—or security—in order to protect a proprietary part of the system. Card system software is written in a low-level machine language for a particular smart card chip and is used to extend or replace basic functions on the smart card.

Table 1.1. Types of smart card software with sample applications.

Software Type	Application	System
Host	Digital signature	Electronic purse
Card	Lottery game	Encryption algorithm

Host and Card Software Integration

Both kinds of smart card software—host software outside the card looking in and card software inside the card looking out—are treated in this book, but they are fundamentally different in their orientation and outlook. Card software focuses on the contents of a particular card. Card software provides computational services for applications in accessing these contents, and protects these contents from many applications which might try to access them incorrectly. Host software, on the other hand, might make use of many different cards. Host software is typically aware of many cardholders and possibly many card issuers as well as many different kinds of cards.

Card software implements the data and process security properties and policies of a particular smart card. For example, a program running on the card might not provide an account number stored on the card unless presented with a correct personal identification number (PIN). Or a program running on the card might compute a digital signature using a private key stored on the smart card, but it would under no condition release the private key itself. Software running on a smart card provides secure, authorized access to the data stored on the smart card. It is only aware of the contents of a particular smart card and entities "out there"—people, computers, terminals, game consoles, set-top boxes, and so on—trying to get at these contents.

Host software connects the smart cards and the users carrying them to larger systems. For example, software running in an automatic teller machine (ATM) uses the smart cards inserted by the bank's customers to identify the customer and to connect the customers with their bank accounts. Or software running in a soda machine verifies that the card inserted into the card reader is a valid cash card and decrements the amount of cash on the card before triggering the release of a can of soda. Host software is aware of many smart cards and tailors its response based on the particular smart card presented.

Unlike most computer software, which relies on supporting services from its surrounding context, smart card software begins with the assumption that the context in which it finds itself is hostile and is not to be trusted. Until presented with convincing evidence to the contrary, smart cards don't trust the hosts they are inserted into and smart card hosts don't trust cards that are inserted into them. A smart card program only trusts itself. Everything outside the program has to prove itself trustworthy before the program will interact with it.

Host Programs

By far, most smart card software will be host software and will be written against existing smart cards, either commodity off-the-shelf smart cards available from smart card manufacturers or smart cards created by major smart card issuers such as bank associations, telecommunication companies, retailers, and national governments. The operating system on these widely distributed smart cards implements a characteristic set of commands—usually 20 or 30—to which the smart card responds. Host software sends commands to the card operating system that executes them on the smart card processor and returns the results. Some examples of commands are "Decrement the amount of money in purse 1 by $1.50," "Authenticate user with PIN 1234," and "Read and return the second record from file 5."

There are more functions you might like a smart card to perform than can fit in the resources of today's smart cards. At this time, there is not an all-purpose smart card operating system or an all-purpose, off-the-shelf smart card. For example, some standard cards are particularly good for payment and loyalty applications, some standard smart cards are particularly good for network and cryptological applications, and some smart cards are particularly good for mobile telephone applications. Also, some off-the-shelf cards offer collections of general-purpose—if low-level—functions. One of the first tasks of a host software programmer is to choose the smart card that will be included in the system.

A host smart card program must accomplish two tasks before it begins to conduct business with a smart card. First, it must ensure that the smart card it is dealing with is authentic. Second, it must convince the card that it is authentic. No business can be conducted before this mutual trust has been established. Actually filling the role the smart card was intended to perform—provide some digital cash or produce a digital signature—is typically only a very small part of the total interaction between a command language host program and the smart card operating system.

High-Level Language Card Programs

Late in 1996, the smart card manufacturer Schlumberger introduced the first off-the-shelf smart card that could accept and run programs written in a high-level programming language, namely Java. Until the advent of the Java Card, the only way to get software onto a smart card was to have it written and loaded into a smart card by a smart card manufacturer. This was a long, tedious, and surprisingly error-prone process. It was also a very expensive process and precluded all but the largest organizations from creating purpose-built smart cards. Some smart card manufacturers used high-level languages such as Forth and C to create card software, but the capability of using these tools to program the card itself was not passed on to the owner of the card and certainly not to the cardholder.

One use of a Java program running on a smart card is to implement a specialized set of commands for use by a host program. The Java program receives commands from the host program, executes them on the card, and returns results just like the operating system in an off-the-shelf smart card. Using this technique, a Java Card can either emulate an existing non-Java Card, extend an existing card with new commands, or become a wholly new card for use by host software.

Theoretically, a Java smart card program could implement any command set whatsoever. However, due to both memory constraints for containing the program on the card and time constraints for running the program, the functionality of the Java program is limited by the functionality of the underlying operating system services it can call on. For example, you likely would not want to perform encryption using very secure keys purely in Java on the card (because of the computational loading on the card), but rather would call on cryptographic functionality in the native card operating system and perhaps in a crypto coprocessor in the microcomputer itself.

Java programs that are both stored and executed on the card have also opened up the possibility of breaking out of the master/slave relationship between host software and card software and thus have enabled new classes of smart card applications. Although the underlying communication channel is still a half-duplex channel (that is, a channel over which either end can send information to the other end, but in only one direction at a time), a smart card programmer can now arrange for the card to control and send commands to the host rather than the other way around.

Assembly Language Card Programs

Although they're the least likely and most demanding of the smart card programming scenarios, there are situations in which a system designer will want to consider extending an existing smart card operating system or creating a new and unique smart card. For example, you might wish to add a new encryption algorithm or mode of communication to a smart card.

To build a custom smart card, an application system designer would typically work with a smart card manufacturer, such as Schlumberger or Gemplus, and possibly a smart card chip manufacturer, such as Motorola, Siemens, or Philips. It might be possible to modify and extend existing operating systems and libraries owned by these manufacturers to incorporate custom features as opposed to creating a completely new operating system for the smart card. Both Schlumberger and Gemplus produce off-the-shelf cards that can be extended after manufacture. Some functions on smart cards, such as communications and card file services, are common to almost all smart cards; the creator of a custom card would not want to reproduce these. A small number of independently developed smart card operating systems are available for licensing.

Creating a custom-built smart card is expensive both in time and in money. Table 1.2 lists some hypothetical cases. You should probably budget at least two years and at least $1 million. Furthermore, it is unlikely that the resultant card will interoperate with any existing host software or systems, so there might be additional time and money expenses on the host side of the system. For all but the largest organizations and applications, assembly language programming of smart cards should be considered only for closed, mission-critical systems where compatibility is not a consideration.

Each smart card chip manufacturer has its own procedures for building a smart card chip containing a custom assembly language program and operating system. These procedures are discussed in detail in Chapter 3, "Some Basic Standards for Smart Cards," as are procedures for getting these chips made into cards.

Table 1.2. Comparison of smart card programming projects.

Software Type	Application Development Time	Typical Program Size	Card Type Used	Difficulty	Expense
Host Application card	6 months	10 KLOC	Off-the-shelf	Medium	Low
	1 year	1 KLOC	Java Card	Medium	Medium
System card	2 years	4 KLOC	Custom	High	High

Smart Card Software Security

Smart card software security is, not surprisingly, based on cryptography. Keys are stored in files on the card and algorithms and protocols are implemented in software on the card. Cryptography is used primarily to authenticate system entities, such as users, cards, and terminals, and to encrypt communications between the smart card and the outside world. The cryptographic functions built into a smart card for its own security requirements may also be used to implement security functionality in other systems. The protections provided by the former obviously enhance the security of the latter.

Before a smart card can provide access to its resources, it must determine with whom it is dealing. Similarly, before it is accepted by other entities, it must be able to prove who it is. Therefore, one of the first tasks a smart card performs when it is activated is to authenticate entities outside itself, primarily the person who inserted it into the terminal and the terminal into which it has been inserted, and to authenticate itself to some or all of these entities.

An authentication procedure may be as simple as demonstrating the possession of a shared secret such as a 4-digit PIN, or it may be as complicated as demonstrating the ability to encode an offered message called a *challenge* with a particular key and algorithm or continuously following a defined transaction protocol. If at any point in an authentication process an entity reveals that it is not who it claims to be, all further communication with the entity is blocked. A record of these failed attempts may be kept on the smart card and after a certain number of failures with no intervening success having been reached, the card may block all further access or destroy itself and its contents completely.

Encryption can be applied to all message traffic to and from the smart card or only to particular messages. If a smart card is communicating with two applications simultaneously, it may be using a different encryption key or technique with each.

Smart card programmers typically do not have to design new authentication or encryption algorithms. Rather, they use the facilities that are built into the smart card. These facilities have been field tested and come with a certain level of assurance of correctness. Designing new algorithms is not easy, and validating the correctness of a new algorithm is probably not a subtask that a smart card application developer wants to assume. Table 1.3 lists a number of cryptographic algorithms which find use in various smart cards.

Table 1.3. Examples of smart card encryption algorithms.

Algorithm	Sample Uses
DES	Communication channels
A3 and A8	GSM mobile telephone
Elliptic curve	Digital signature
TSA7	Health records
RSA	Digital signature

Smart Card Operating Systems

Smart card operating systems are not operating systems in the sense that today's mainstream programmers and software developers think of them. Smart card operating systems certainly don't have the functionality of Windows or UNIX or even DOS. Rather, they are more like pre-DOS collections of on-card commands to which the smart card responds.

The basic relationship between a smart card terminal, such as a personal computer into which a smart card is inserted, and the smart card itself is one of master and slave. The terminal sends a command to the smart card, the smart card executes the command, returns the result if any to the terminal, and waits for another command.

In addition to describing the physical characteristics of a smart card and the detailed formats and syntaxes of these commands and the results they return, smart card standards such as ISO 7816 and CEN 726 also describe a wide range of commands that conforming smart cards can implement. Most smart card manufacturers offer off-the-shelf smart cards with operating systems that implement some or all of these standard commands, together with manufacturer-specific extensions and additions. Table 1.4 lists a few examples.

Table 1.4. Examples of operating systems for off-the-shelf smart cards.

Card Type	Manufacturer	Maximum memory size	Extensible
Multiflex	Schlumberger	8 KB	Yes
MPCOS64K	Gemplus	8 KB	Yes
USCO48	US3	8 KB	No
OC100	Bull CP8	8 KB	No
I006.1	Orga	4 KB	No

Smart Card File Systems

Most smart card operating systems support a modest file system based on the ISO 7816 smart card standard. Because a smart card has no peripherals, a smart card file is really just a contiguous block of smart card memory. A smart card file system is a singly rooted directory-based hierarchical file system in which files can have long alphanumeric names, short numeric names, and relative names.

You usually can't extend or expand an already allocated file, so you have to create files at the maximum size they are expected to be. Furthermore, there is no notion of garbage collection or compacting in a smart card file system. So, for example, if file A and file B are created in this order in the same file system directory, and then file A is deleted, the space occupied by A is lost until file B is also deleted. If file B is deleted because it was the last file created, the space occupied by file B is reclaimed and can be reused by file C, for example.

Smart card operating systems support the usual set of file operations such as create, delete, read, write, and update on all files. In addition, operations are supported on particular kinds of files. *Linear files*, for example, consist of a series of fixed-size records that can be accessed by record number or read sequentially using read next and read previous operations. Furthermore, some smart card operating systems support a limited form of seek on linear files. *Cyclic files* are linear files that cycle back to the first record when the next record after the last record is read or written. Purse files are an example of an application-specific file type supported by some smart card operating systems. *Purse files* are cyclic files, each of whose records contain the log of an electronic purse transaction. Finally, *transparent files* are single undifferentiated blocks of smart card memory that the application program can structure any way it pleases.

Associated with each file on a smart card is an access control list. This list records what operations, if any, each card identity is authorized to perform on the file. For example, identity A may be able to read a particular file but not update it, whereas identity B may be able to read, write, and even alter the access control list on the file.

One of the key design tasks of a smart card programmer in building a new smart card application is defining and laying out the files that the application will expect to find on the smart card; for example, see Table 1.5. Size and usage of smart card real estate is a key concern, but the names of the files, their hierarchical relationship to one another, and their access authorizations must also be thought through.

Table 1.5. Examples of smart card file types.

Type	Special Operations	Sample Uses
Linear	Seek	Credit card account table
Cyclic	Read next, read previous	Transaction log
Purse	Debit with certificate	Loyalty points, e-cash
Transparent	Read and write binary	Picture
SIM file	Encrypt, decrypt	Cellular telephone

Smart Card Communications

The typical single-chip computer on a smart card is able to transmit and receive data at speeds up to 115,200 bits per second (bps), but most smart card terminals drive smart cards at 9,600 bps for contact smart cards and 7,800 bps for contactless smart cards. Because only a small amount of data is being transmitted and because the channel can be quite noisy, reliable communication is more important than high-speed communication.

The communications pathway to and from a smart card is *half-duplex*; that is, data is either flowing from the terminal to the smart card or from the smart card to the terminal, but not both at the same time. The result is that the smart card and the terminal have to be synchronized and always agree on whose turn it is to talk and whose turn it is to listen. If both the terminal and the card transmit at the same time, data will be lost. If they both believe it is their turn to listen, the system enters a deadlock situation and nothing further happens.

Data received by and transmitted from a smart card is stored in a buffer in the smart card's very limited random access memory. Therefore, only relatively small packets—10 to 100 bytes—of information are moved in each message. Although the ISO and CEN standards describe in detail the format and coding of these messages, nothing prevents the smart card programmer from designing messages specifically tuned to his or her application. You must ensure that other surrounding systems don't make assumptions about the messages based on the ISO or CEN standards (which you'll learn more about in Chapter 3).

Smart Card Hardware

The computer on a smart card is a single integrated circuit chip that includes the central processing unit (CPU), the memory system, and the input/output lines. A single chip is used in order to make tapping into information flows inside the computer more difficult. If more than one chip were used to implement the smart card computer, the connections between the chips would be obvious points of attack.

The Smart Card Memory System

Smart cards have a memory architecture that will be unfamiliar, if not downright bizarre, to most mainstream programmers. Programmers typically think in terms of having available large amounts of homogeneous random access memory (RAM) that is freely available for reading and writing. This is definitely not the case on a smart card. There are, in fact, three kinds of memory on a smart card: read-only memory (ROM), nonvolatile memory (NVM), and a relatively tiny amount of random access memory (RAM).

Read-only memory is where the smart card operating system is stored and is of interest only to assembly language programmers. General-purpose smart cards have between 8 KB and 32 KB of ROM. Here, one finds various utility routines such as those for doing communication and for maintaining an on-card file system along with encryption routines and special-purpose arithmetic routines. Code and data are placed in read-only memory when the card is manufactured and cannot be changed; this information is hardwired into the card.

NVM is where the card's variable data—such as account numbers, number of loyalty points, or amount of e-cash—is stored. NVM can be read and written by application programs, but it doesn't act like and cannot be used like RAM. NVM gets its name from the fact that it retains its contents when power is removed from the card; data written to NVM, if not overwritten, will last 10 years. NVM presents two problems:

- *Slowness.* It generally takes 3–10 milliseconds to write data into NVM.
- *Data loss.* NVM wears out after it has been written to a number of times (around 100,000 times).

The typical programmer is not familiar with either of these two problems, but must take them into account when writing smart card software.

There is some familiar RAM on a smart card, but not very much—usually only 1,000 bytes or less. This is unquestionably the most precious resource on the smart card from the card software developer's point of view. Even when using a high-level language on the smart card, the programmer is acutely aware of the need to economize on the use of temporary variables. Furthermore, the RAM is not only used by the programmer's application, but also by all the utility routines, so a programmer has to be aware not only of how much RAM he is using, but how much is needed by the routines he calls.

The Smart Card Central Processing Unit

The central processing unit in a smart card chip is an 8-bit microcontroller, typically using the Motorola 6805 or Intel 8051 instruction set. Hitachi's H8 smart card chip is a notable exception. These instruction sets have the usual complement of memory and register manipulations, addressing modes, and input/output operations. Some chip manufacturers have extended these basic instruction sets with additional instructions that are of particular use on smart cards. Smart card CPUs execute machine instructions at the rate of about 400,000 instructions per second (400 KIP), although speeds of up to 1 million instructions per second (1 MIP) are becoming available on the latest chips.

The demand for stronger encryption in smart cards has outstripped the ability of software for these modest computers to generate results in a reasonable amount of time. Typically 1 to 3 seconds is all that a transaction involving a smart card should take; however, a 1024-bit key RSA encryption can take 10–20 seconds on a typical smart card processor. As a result, some smart card chips include coprocessors to accelerate specifically the computations done in strong encryption.

A smart card CPU will not necessarily execute code from all parts of the smart card memory system. Most smart card chips, for example, will not execute code stored in RAM. Furthermore, some chips make it possible to reconfigure sections of NVM so that a program loaded into NVM cannot be overwritten (essentially turning the NVM into ROM) or so that the CPU won't take instructions and therefore execute code from this part of memory.

Smart Card Input/Output

The input/output channel on a smart card is a unidirectional serial channel. This means that it passes data 1 bit and hence 1 byte at a time, and that data can flow in only one direction at a time. The smart card hardware can handle data at up to 115,200 bps, but smart card readers typically communicate with the card at speeds far below this.

The communication protocol between the host and the smart card is based on a master (host) and slave (smart card) relationship. The host sends commands to the card and listens for a reply. The smart card never sends data to the host except in response to a command from the host.

Smart card operating systems support either character or block transfers, but usually this level of detail is hidden from the smart card programmer.

Smart Card System Design

Most smart card programming consists of writing programs on a host computer that send commands to and receive results from predefined or application-specific smart cards. These applications read data from and write data to the smart card and perhaps make use of the modest computing powers of the processor on the smart card. Smart cards in these applications are typically secure stores for data pertaining to the individual bringing the card to the system, such as personal identification data.

In situations where no off-the-shelf card contains all the functionality needed by the application, the programmer may be able to extend the capabilities of an off-the-shelf card by writing software that runs on the card itself. This software may implement special-purpose or higher-level functions on the card that is a combination of existing operating system functions, or it may provide additional protections for the data stored on the card.

Finally, there may be situations where the operating system capabilities of an existing smart card need to be extended, or where a new and unique smart card needs to be manufactured. Examples of such situations include a closed system application where cost or a particularly high level of security is a critical factor, or where a particular encryption algorithm is needed to connect the smart card to an existing host system. In these situations, smart card programmers write new operating system software for smart cards partially or completely in the assembly language of the processor on the smart card.

Regardless of the type of software being written, the smart card programmer must be constantly aware of the two central concerns of smart card software: data security and data integrity.

Data Security

Data security means simply providing data only to those who are authorized to receive it. It requires that neither data values nor even information about these data values are revealed to unauthorized parties or systems. Wherever data is stored and whenever data is moved, the smart card programmer must ensure that this requirement is satisfied.

Although there are many methods for obscuring data, data security doesn't make use of data encryption as much as it does the notion of an authorized entity. One of the most obvious attacks on a smart card is for an unauthorized entity to become authenticated as an authorized entity, then use the provided facilities of the card to access data. Because keys, passwords, and PINs stored on the card are all used to authenticate entities, particular care must be exercised in protecting this kind of data.

One way to ensure the data security of a smart card is to control physical access to the card. This technique is often used early in the life cycle of the card. For example, cards are manufactured under tight physical security procedures and shipped to the card customer under equally tight procedures. The keys used to protect the card during transit from manufacturer to customer are called *transport keys* and they are given only to the customer. Using the transport keys, a smart card customer can access all files on the smart card and set up a particular file system and access authorization scheme. Transport keys are like a superuser password for the card and are typically deleted from the card by the customer after they have been used to configure the card for the customer's application.

Data Integrity

Data integrity requires that all parties to a smart card transaction agree on the state of the data in the transaction. If vending machine software intends to charge a smart card electronic purse $1 for a can of soda, then at the end of the transaction, there should be $1 less on the smart card, $1 more in the vending machine, and a soda in the hands of the cardholder. Any variation—$1 added but $1 not subtracted, $1 transferred, but no soda vended, and so on—is a violation of data integrity.

The primary source of breakdowns in data integrity are system failures in the middle of processing a transaction. Such failures can be accidental—for example, when a communication line from an ATM goes down—or deliberate—for example, when the cardholder prematurely removes the smart card from the reader. Programming to ensure data integrity means both guarding against its loss and detecting and repairing its loss when it occurs.

Smart Card System Architectures

Because it carries neither power nor clock, there is no such thing as a stand-alone smart card. All smart cards are integrated into larger systems that themselves typically contain additional computers and data stores. For example, an ATM system will include a computer in each ATM machine, transaction servers to concentrate requests coming from many ATMs, and database machines that contain identity and account information. In fact, it is in large, distributed, multiparty systems where smart cards play one of their unique roles as a secure identity token.

There have been a number of efforts to specify and de facto standardize smart card systems in addition to just smart cards themselves. Three bank card associations, Europay, MasterCard, and Visa, teamed up in early 1996 to produce EMV'96, which specifies an overall system architecture for debit and credit card applications of smart cards. Late in 1996, a work group—headed by Microsoft and including Schlumberger, Bull CP8 Transac, Hewlett-Packard, and Siemens Nixdorf Information Systems—produced a specification called PC/SC (personal computer/smart card) for connecting smart cards to personal computers and for surfacing application program interfaces to smart cards on personal computers. Next, in early 1997, IBM published a specification called the Open Smart Card Architecture and the Open Group published an architecture for Personal Security Modules. Finally, version 2.0 of the Secure Electronic Transactions (SET) specification published by MasterCard and Visa in the summer of 1997 contains yet another smart card system architecture specification based on the EMV specification and specifically tailored for credit and debit applications of smart cards.

One of the design challenges for a smart card programmer is distributing system functionality among the multiple computers that deal with a smart card and with which a smart card must deal. Deciding what functions go on the smart card itself, what functions go on the machine into which the smart card is inserted, and what functions go on the various machines upstream from the terminal machine comprises the heart of the application design process. Not surprisingly, these are not solely technical decisions; they directly affect both the business interests and security concerns of owners of all these various computers.

Summary

You'll get the most out of this book if you read it with a smart card project in mind. It does not have to be a large project, and in fact it is probably better if it isn't. Having in mind something you would like to do with smart cards lets you connect the many details contained in this book to a familiar situation and thus give them life and an in-context interpretation. Of course, actually implementing the project as you read the book is the best approach of all.

CHAPTER 2

PHYSICAL CHARACTERISTICS OF SMART CARDS

Smart cards present a variety of faces, depending primarily on the type of integrated circuit chip embedded in the plastic card and the physical form of the connection mechanism between the card and the reader. They can be very inexpensive tokens for financial transactions such as credit cards, telephone calling tokens, or loyalty tokens from a variety of businesses. They can be access tokens for getting through locked doors, riding on a train, or driving an automobile on a toll-road. They can function as identity tokens for logging in to a computer system or accessing a World Wide Web server with an authenticated identity. Three such variants are of particular interest:

- Cards with surface contacts leading to a memory-only integrated circuit chip
- Cards with an electromagnetic connection to a microprocessor-integrated circuit chip
- Cards with surface contacts leading to a microprocessor-integrated circuit chip.

The very earliest smart cards were memory cards containing an integrated circuit chip comprised of only nonvolatile memory and the necessary circuitry to read and write that memory. Today, such cards still constitute the largest number of smart cards in use. These cards are relatively inexpensive and provide modest security for a variety of applications.

A *memory card*, as its name implies, is a card that contains an embedded integrated circuit chip providing nonvolatile memory for storing information in a permanent or semi-permanent fashion. The circuitry of the smart card exposes, through a standard electrical connector, the control lines for addressing selected memory locations as well as for reading and writing those memory locations through the electrical connectors on the face of the card. There is no on-board processor to support a high-level communications protocol between the reader and the card. Rather, memory cards use a synchronous communication mechanism between the reader and the card. Essentially, the communication channel is always under the direct control of the reader side. The card circuitry responds in a direct (synchronous) way to the very low-level commands issued by the reader for addressing memory locations and for reading from or writing to the selected locations. In some recent memory cards, security enhancements have been incorporated through the provision of memory addressing circuitry within the chip that requires a shared secret between the terminal (which is writing to the card chip) and the chip itself. These are often called *logic cards*.

A *contactless card* has an integrated circuit chip embedded within the card; however, it makes use of an electromagnetic signal to facilitate communication between the card and the reader. With these cards, the power necessary to run the chip on the card is transmitted at microwave frequencies from the reader into the card. The separation allowed between the reader and the card is quite small—on the order of a few millimeters. However, these cards offer a greater ease of use than cards that must be inserted into a reader. This ease of use can be mitigated by other factors.

With the current state of technology, the data transfer rate between the reader and the contactless card may be restricted by the power levels that can be achieved in the card; that is, for such cards without an internal power source (for example, a battery), the power to run the on-card processor must be derived from a signal transmitted to the card from the reader. The power levels achieved typically allow only a very small separation (a few millimeters) between the card and the reader. Further, a feedback mechanism from the reader to the card through which card holder verification is done is a bit more awkward with the contactless card. Consequently, these cards are most popular for uses where the possession of the card is deemed to be adequate authorization for card use.

Except for the physical mechanism used to transfer information between the reader and the card, contactless and contact-based cards are very similar in overall architecture. This book focuses mainly on smart cards that make use of electrical connections between the cards and the readers. This type of microprocessor-based smart card combines all the necessary ingredients for an enhanced-security computing platform. It integrates both memory and a central processing unit into a single integrated circuit chip. This minimizes the opportunity to intercept well-defined electrical signal patterns moving between processor and memory elements. Keep in mind that the security resulting from this integrated packaging is not infallible. The smart card is tamper-resistant, not tamper-proof.

What's in the Card

The impetus for creating smart cards was the need for secure tokens that can contain information and can provide a secure platform for certain processing activities. These capabilities were greatly facilitated with an innovative packaging approach for the principal elements of a computer system, as illustrated in Figure 2.1. Specifically, all the basic components of a computer system are incorporated into a single integrated circuit chip. This means that the physical connections between these components are embedded within a monolithic (silicon) structure. This, in turn, means that it is difficult for an observer to intercept signals passing between these components (within the chip). The net result is a more secure computer system than is normally achieved with macroscopic physical connections between components.

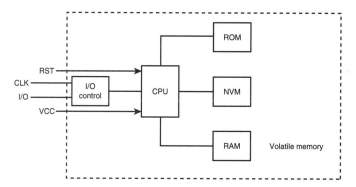

Figure 2.1. *Elements of a smart card computer system.*

Integrated Circuit Chips

The smart card packaging approach consists of putting the central processing unit, all the memory, and the I/O electronics into the same integrated circuit chip, rather than presenting them in the form of various chips which are then tied together through some type of electrical connections. Why is this simple packaging approach so profound? Because it provides all the necessary capabilities in a very small physical package and it conceals the interconnections between the various computer elements inside the chip itself, thereby enhancing the security of what's going on (or what's stored) in the computer.

Once the elements of the computer are integrated into a single chip, it becomes very difficult for an outside observer to intercept signals flowing among the various elements and to subsequently discern the information content of those signals. The connection to the outside world through which information flows is a simple I/O port that can be guarded, to a large extent, by the processor included within the chip. This is done through the use of high-level telecommunications protocols through which the chip's processor element filters all information that is passed to or from the other components of the chip. Through these protocols, it is possible to require authentication of the identity of the reader-side program that is communicating with the computer on the smart card. In this manner, the smart card can protect itself by communicating only with entities that can prove who they are and that the smart card's computer trusts. (Chapter 9, "Smart Cards and Security," includes further discussion on what is meant by *trust* and *security*.)

In addition to enhancing the security of the smart card, the integrated circuit chip packaging also provides a small unit that is amenable to being embedded in a credit card–sized card which can be carried by the card bearer. When embedded in the plastic card and carried, for example, in a person's wallet, the chip is subject to a variety of physical forces. The card is bent and flexed and may be subjected to sudden shocks. In typical electronic equipment, in which components are tied together through macroscopic electrical wiring or even conducting lines on a printed circuit board, this physical environment is an excellent recipe for many failures. When all the elements are packaged in a single chip, however, the stresses tend to be applied equally to all the elements. So, if the chip itself can hold together, then the components will tend to operate successfully. Empirical evidence indicates that when chips are reduced to a size of approximately 25 square millimeters (in roughly a square configuration), they are able to withstand the day-to-day stresses encountered through normal credit card–type uses.

Achieving the small size of the chip is dependent on several criteria:

- The resolution of the technology used for the chip, which is often characterized by "feature size" (for example, the size of a single transistor element within the chip) in microns
- The width of the internal bus of the processor; that is, is it 8 bits, 16 bits, 32 bits, or 64 bits?
- The type of memory used
- Auxiliary elements (such as power line frequency, voltage filters, and memory-mapping registers) included in the chip for security or functionality reasons

Size

The small size needed for chip features requires leading-edge technology. However, in order for chips to be inexpensive and reliable, we often need to turn to older, more mature technologies.

Width

The width of the internal bus structure indicates the number of memory address lines running between components within a chip; that is, width is generally indicative of the number of bits in individually addressable sections of memory. Minimizing chip size generally tends to call for selection of fewer address lines; therefore, most smart card chips are currently based on 8-bit microprocessors. These microprocessors also tend to be the older and more mature technologies.

Memory

The type of memory used in smart card chips brings a very interesting wrinkle. The implementation technologies used for chip memories vary greatly in the size of individual memory cells. The smallest memory element is read-only memory (ROM). This type of memory, as the name implies, can be read by typical computer elements, but it requires very special equipment in order to write information into the memory. In fact, the writing of ROM can be incorporated very early into the chip fabrication process itself. This technique tends to enhance the security of the chip, since it is difficult to examine the contents of the ROM without destroying the chip—even with very expensive probing equipment. So this type of memory is useful for permanently encoding stored programs for the smart card, but it is useless for storage of dynamic information that needs to be changed during the normal use of the card.

Significantly larger is the electrically erasable and programmable read-only memory (EEPROM). The contents of this type of memory in a smart card chip can actually be modified during normal use of the card. Hence, programs or data can be stored in EEPROM during normal operation of the card and then read back by applications that are using the card. The electrical characteristics of EEPROM memory are such that it can only be erased and then reprogrammed a finite (but reasonably large) number of times, generally around 100,000 times. While somewhat limited, techniques have evolved which make this type of memory quite useful for typical smart card uses. EEPROM memory cells tend to be close to a factor of four larger than ROM memory cells. EEPROM, like ROM, does have the nice characteristic of being nonvolatile memory; that is, the information content of the memory is unchanged when the power to the memory is turned off. Information content is preserved across power-up and power-down cycles on the smart card chip.

Larger still is a memory type known as random access memory (RAM). This is the type of memory used in typical computer systems, such as a desktop PC. Information can be written and erased many times in this type of memory. In the smart card chip, however, a RAM memory cell is approximately four times larger than an EEPROM memory cell. RAM is also volatile memory; that is, the contents of the memory are lost when power is removed from the memory cell. So information in RAM is not preserved across a power-down and power-up cycle on a smart card. RAM is, nevertheless, essential for certain operations in smart card applications; in particular, it requires much less time for RAM locations to be read or written by the chip's processor unit. This can be extremely important when the smart card is interacting with a PC application in which the timing of responses from the card to the PC are important; this is often the case in the mobile telecommunications area (that is, smart card–based cellular telephones).

The net result is that smart card chips tend to make use of varying amounts of each memory type, depending on the specific application for which the smart card is to be used. The most powerful chips used in smart cards today have RAM sizes in the 256-byte to 1-KB range, ROM sizes in the 16-KB to 32-KB range, and EEPROM sizes in the 1-KB to 16-KB range.

Coprocessors

A typical smart card processor is an 8-bit microprocessor. Such a processor is capable of manipulating only 1 byte of information at a time. This manifests itself in the support of 8-bit integer arithmetic as the primary computational facility of the computer. Handling larger-integer arithmetic or floating-point arithmetic operations requires significant additional programming beyond the basic instruction set of the processor. This presents something of a problem when you need to support public key cryptography on a smart card chip.

Public key cryptography is predicated on the use of integer arithmetic on a scale which severely taxes the capabilities of a typical smart card processor. Performing encryption or decryption operations can be extremely time-consuming, taking several seconds or even minutes. Since these delays are not acceptable given the time it should take to conduct a typical transaction, enhancements to smart card processors are needed. This enhancement has been accomplished by adding to the chip a second processor that is capable of enhanced performance for selected integer arithmetic operations, such as fast integer multiply operations. This greatly speeds up the public key cryptography operations; however, it affects the overall size of the chip (slightly) and the cost of the chip (more significantly).

Security Features

Physical security of information stored in a smart card starts with the combination of computer memory and processor in the same small package. It is difficult, though not impossible, to physically examine the contents of memory cells within the chip. It is also difficult, though not impossible, to intercept the electrical signals passing between the processor and memory or between processor elements during selected computations. To examine or intercept such information requires the use of fairly expensive equipment and unfettered access to the smart card itself, usually without the smart card's owner being aware of it.

Security features are sometimes enhanced by randomizing the sequence of memory cells accessed by the processor. That is, the address lines for various memory cells don't proceed in a linear sequence, but rather are varied from one cell to the next through some complex algorithm. The net result is that an external observer is less likely to be able to discern any information about where data is stored or how it is being used by simply watching the sequencing of access to individual memory cells.

As the use of smart cards has grown, the number of attempts to thwart the security features of smart cards has grown. Several techniques to coax information out of a card have been identified. Some of these involve manipulation of the power supplied to a card. Defenses against these techniques have been developed as well. Some chips have additional sensors which monitor characteristics of the power supplied to the chip. This information can be used by programs within the chip and allow it to lock down the card when it detects that it is under attack. In extreme circumstances, the card can destroy sensitive information in such cases in order to prevent it from being extracted by the attacker.

The Manufacturing Process

The manufacturing of smart cards comprises a number of distinct steps:

1. *Fabrication of the chip, or many chips in the form of a wafer.* Several thousand integrated circuit chips are manufactured at a time in the form of silicon wafers. An individual chip for a smart card is approximately 25 square millimeters, or about 5 millimeters on a side. The template for the circuitry on a chip is repeated many thousands of times to overlay a silicon wafer approximately 4 inches in diameter. Such a wafer might routinely contain 3,000 to 4,000 chips when completed. The actual fabrication of the chips on the silicon wafer is done through a highly refined process of vacuum deposition of extremely pure semiconductor material on the silicon substrate (see Figure 2.2).

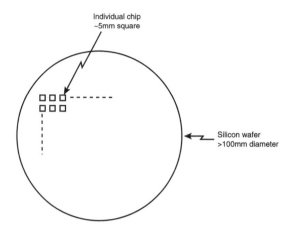

Figure 2.2. *Chip fabrication.*

2. *Packaging of individual chips for insertion into a card.* Once a wafer is completed, each individual chip on the wafer must be tested to make sure that it is operable. Each good chip is identified by a physical marking in preparation for sawing the wafer into many thousands of pieces (that is, one chip in each piece). Once the chips are segmented, an electrical connector which is larger than the chip itself is attached. Very tiny electrical connectors (wires) link various areas on this connector with specific pins on the chip itself. The resulting configuration is referred to as a *module.* Figure 2.3 illustrates the components of a module, including the micro-electronic connections between the connector and the chip.

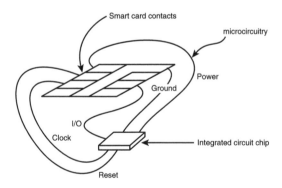

Figure 2.3. *Elements of a smart card module.*

3. *Fabrication of the card.* The card itself is constructed out of polyvinyl chloride or a similar material. Both the chemical characteristics and the dimensions of the card and its associated tolerances are stringently regulated by international

standards (which are further discussed in Chapter 3, "Some Basic Standards for Smart Cards"). The card material is produced in a large, flat sheet of the prescribed thickness. For many types of mass-produced cards, these sheets are then printed. Individual cards are then punched from this flat sheet and the edges of each card are ground to a smooth finish.

4. *Insertion of the chip into the card.* Once the module and card are prepared, the two are brought together during an insertion operation. A hole is made in the card, and the module is glued into it. This hole is typically produced either through a milling operation or by melting the material and pushing the module directly into it (see Figure 2.4).

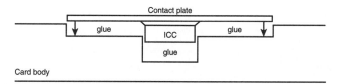

Figure 2.4. *Module insertion in card body.*

5. *Pre-personalization.* Once the module is inserted into the card, most smart card applications require that certain programs or data files be installed on each chip (card) before the card can be personalized and given to a specific cardholder. This general preparation of software or files on the card is done through an operation called *pre-personalization*, which is done through the I/O connectors on the surface of the card and hence can proceed only at the speed supported by that interface.

6. *Personalization.* The personalization procedure involves putting information such as names and account numbers into the chip on the card. This also usually entails writing a personal identification number (PIN) on the card which the cardholder can then use to confirm his or her identity to the card. The personalization procedure usually involves physical manipulation of the card as well; that is, pictures, names, and address information is often printed on the card. In addition, some information, such as account numbers, may be embossed on the card to allow physical transference of that information onto other media (for example, printing a paper receipt of a credit card transaction).

7. *Printing of the card.* The printing of graphics and text on a smart card is an extremely important feature. The appearance of the card generally reflects both aesthetically and financially on the issuer of the card. Branding

information such as corporate symbols and logos builds name recognition for the issuer and has significant advertising value. When a card is used as a personal identification card, a person's picture along with name and address information is often printed on the card. For many cards supporting financial transactions, issuers are often concerned with the threat of counterfeiting of their cards; they will sometimes make use of anti-counterfeiting mechanisms such as holograms printed on the face of the card as a safeguard, much as is found in thwarting the counterfeiting of currency.

Depending on the information to be placed on a card, a variety of printing processes may be used. For cards that will have exactly the same graphics on every card, such as telephone cards, transit tokens, and the like, the printing step is often done prior to the insertion of the integrated circuit chip in the card. In this case, the cards are formed from large plastic sheets. The printing is done on the sheet, prior to cutting the individual cards from the sheet. Following the printing operation, the individual cards are stamped from the sheet and their edges are smoothed to conform to ISO specifications, which is examined in a bit more detail in Chapter 3.

8. *Initialization of the program and program information on the chip in the card.*

Smart Card Application Software Development

Applications that make use of smart cards typically comprise software that runs on a reader-side computer (a PC or PC class computer), on the smart card itself, and perhaps even on other computers widely distributed within a local area or wide area network. Development of such applications require special considerations to fully incorporate the smart card, both with respect to development of software to go on the card itself and in testing the smart card in the full application. Some of the elements of this application software development that are specific to the smart card component are described in the following sections.

Mask Development

Programs stored into the chip on a smart card are often referred to as *masks*. The term derives from the fact that the software is actually reduced to a bit pattern which is actually masked onto the silicon components (the ROM) of the chip during the fabrication process. If the program is to be stored into ROM as part of the fabrication process of the chip itself, then this is generally referred to as a *hard mask*.

On a chip, software can also be stored in, and executed from, EEPROM. In this case, the software can actually be written to the chip after the card manufacturing process is completed. This type of software is often referred to as a *soft mask* in the smart card domain. For other domains—that is, in other areas where embedded microprocessor units are used—this would often be referred to as *firmware*.

Code Development

The available memory on a smart card chip (for program storage) is typically quite small; at least it's small compared with that of PC platforms. Further, the deployment of large numbers of cards (often millions of them) tends to push in the direction of providing the absolute minimum amount of memory to support the desired application. Consequently, the development of software for smart cards has tended in the direction of highly specialized software for each new application on a chip. The idea of a general operating system on a chip that could readily support a wide variety of applications is a relatively new concept for smart cards.

Chip Simulators

Because software for a smart card is usually loaded onto the chip during the manufacturing process, the process for writing and then checking software (debugging) can be very long. In order to minimize this, most chip manufacturers provide software simulators for their chips. This allows the software developer to create a full complement of software for a chip and check it out via the simulator prior to fabricating a set of chips with the software included.

While a chip simulator improves the software development process, it still leaves many aspects of the software unchecked. For example, it is very difficult to check the timing of various operations through a simulator. As will be seen in Chapter 3, many aspects of the communication between the reader side and the card side of a smart card application are critical, depending on the timing of various transactions between the reader and the card. This aspect of the software's execution can usually not be tested with the simulators provided for most chips.

Chip Emulators

To improve the testing environment, but without requiring the actual fabrication of chips with embedded application software, most chip manufacturers provide hardware emulators for their chips. With an emulator, a variant of memory is provided which can be accessed both by a computer being used for the software development environment and by the processor of a chip of the form to be deployed on the smart card. This emulator allows the software developer to write code and load it into this sharable memory. The code thus loaded can then be executed by the processor in the emulator. In this way, the code is being run in an environment much like that in the final smart card; the timing of operations is much closer to that which will be found on the smart card itself.

Protocol Analyzers

The communication between the reader and the smart card occurs through a half-duplex communication channel, much like is found with a typical PC connection to a local or wide area network. In this environment, it is very useful to be able to monitor the bits traveling across the interface lines between the reader and the card. Just as in the case with a local area or wide area network connection, this can be accomplished with a protocol analyzer.

Interface Devices (Readers)

A smart card contains no independent power source or clock signal to drive its embedded processor. Consequently, it must be plugged in to a device that can provide it with both power and clock. This device is formally referred to as an *interface device* and is often called a *reader* or a *terminal*. The term *reader* is generally used when the reader is providing a connection between a smart card and another computer system such as a PC. The term *terminal* is used when the reader is actually an integral part of a specialized computer system which handles smart cards as an integral component. A general form of such a system is a point-of-sale terminal such as is found in a restaurant that accepts credit cards for financial transactions.

Summary

Smart cards are plastic cards with embedded integrated circuit chips. These integrated circuit chips are highly specialized in that they contain memory and processor elements within a single integrated circuit chip. This configuration enhances the security of the chip as it shields the relatively low-level electrical signals between elements from view by the outside world. The physical environment in which smart cards are used requires that the chips be physically small. This physical limitation greatly restricts the amount of memory that can be incorporated on the chip.

CHAPTER 3

SOME BASIC STANDARDS FOR SMART CARDS

The precursors to today's smart cards were the plastic credit cards used as identity tokens in the retail financial marketplace. These tokens, when introduced, represented a significant improvement in the ability of merchants to accept payment in an abstract form (essentially "on credit") from customers whose identity they could not personally vouch for. The credit card represented (and actually still does) a certification of identity and financial situation from an issuer functioning as a trust broker. (This concept of a trust broker is examined in more detail in Chapter 9, "Smart Cards and Security.")

In the trust infrastructure provided by the credit card—although a merchant might not be inclined to extend credit to an unknown customer—it was reasonable for the merchant to trust the issuer of a credit card, in no small part because of the known financial strength of the issuer and financial agreements entered into when the merchant became "certified" to accept the (credit card) tokens of the card issuer.

As the convenience of credit cards became established and their use more accepted, it became highly desirable to achieve an unprecedented level of interoperability among cards from different issuers and transaction equipment from a variety of vendors in merchants' stores around the world.

The start of the journey toward worldwide interoperability lay in the establishment of international standards regarding first the cards themselves, then the equipment that would work with them and with the environments in which they would be used. The venue of choice for establishing such standards is the International Standards Organization (ISO). In some fields of technical activity, the International Electrotechnical Commission (IEC) collaborates with the ISO in the development of standards. Similarly, in the United States, the American National Standards Institute (ANSI) functions as a primary standards setting body. In the following discussion, some standards are ISO standards and others are joint ISO/IEC/ANSI standards accepted by all three bodies.

This chapter and Chapter 4, "Smart Card Commands," review some of the foundation standards on which smart cards are based. This is not an exhaustive review of the standards affecting smart cards or of the individual standards addressed. Rather, it is a moderately detailed overview of the characteristics of smart cards that are affected by standard specifications; it *will* give you the tools you need to identify the specific standards you can consult for a complete understanding.

ISO, IEC, and ANSI Standards for Cards

ISO/IEC (and in some instances, ANSI) standards have been established to fully describe plastic identification cards. The various standards have evolved over time and are found in a variety of ISO/IEC/ANSI classifications, the more pertinent of which are listed in Table 3.1.

Table 3.1. ISO standards pertaining to smart cards.

ISO/IEC Standard	Title
ISO/IEC 7810—1995-08-15	Identification Cards—Physical Characteristics
ANSI/ISO/IEC 7811-1—1995: Part 1	Identification Cards—Recording Technique Embossing
ANSI/ISO/IEC 7811-2—1995: Part 2	Identification Cards—Recording Technique Magnetic Stripe
ANSI/ISO/IEC 7811-3—1995: Part 3	Identification Cards—Recording Technique Location of Embossed Characters on ID-1 Cards
ANSI/ISO/IEC 7811-4—1995: Part 4	Identification Cards—Recording Technique Location of Read-Only Magnetic Tracks—Tracks 1 and 2
ANSI/ISO/IEC 7811-5—1995: Part 5	Identification Cards—Recording Technique Location of Read-Write Magnetic Tracks—Track 3
ANSI/ISO/IEC 7812-1—1993	Identification Cards—Identification of Issuers Part 1: Numbering System
ANSI/ISO/IEC 7813—1995	Identification Cards—Financial Transaction Cards

ISO/IEC Standard	Title
ISO 7816-1	Identification Cards—Integrated Circuit(s) Cards with Contacts—Physical Characteristics
ISO 7816-2	Identification Cards—Integrated Circuit(s) Cards with Contacts — Dimensions and Location of the Contacts
ISO 7816-3	Identification Cards—Integrated Circuit(s) Cards with Contacts—Electronic Signals and Transmission Protocols
ISO 7816-3 Amendment 1	Protocol type T=1, Asynchronous Half Duplex Block Transmission Protocol
ISO 7816-3 Amendment 2	Revision of Protocol Type Selection
ISO 7816-4	Identification Cards—Integrated Circuit(s) Cards with Contacts—Interindustry Commands for Interchange
ISO 7816-5	Identification Cards—Integrated Circuit(s) Cards with Contacts—Number System and Registration Procedure for Application Identifiers
ISO 7816-6	Identification Cards—Integrated Circuit(s) Cards with Contacts—Interindustry Data Elements
ISO 1177—1985	Information Processing—Character Structure for Start/Stop and Synchronous Character-Oriented Transmission

Physical Characteristics of Identification Cards

A seminal specification that ultimately leads to smart cards is ISO/IEC 7810: Identification Cards—Physical Characteristics. This standard defines nominal physical characteristics for three types of identification cards, labeled ID-1, ID-2, and ID-3. Card type ID-1 deals with the generally accepted size and shape of a "credit card" or "smart card" and is the primary focus of this discussion. The ID-2 and ID-3 card types are simply larger sizes, but with the same physical characteristics as ID-1 card types.

The basic function of an ID-1 identification card is to contain information in a visual, tactile, and electronic form that identifies its bearer and that may support transactions the card is to enable. Visual information may be presented through artwork, personal pictures, anticounterfeiting mechanisms (such as holograms), or machine-readable mechanisms (such as barcodes). Embossing is used to convey information in a tactile form suitable for creating impressions of characters on other media. This is a capability often used in transactions based on such identification cards. Information may be conveyed in electronic form through two mechanisms—magnetic stripes, which are prevalent on credit cards, and an embedded integrated circuit chip, which is the defining characteristic of smart cards.

The elements of an ID-1 identification card include

- The card backing (the plastic)
- Optional embossing areas on which alphanumeric information may be stamped
- An optional area to which a magnetic stripe may be attached

Information can then be magnetically encoded on the magnetic stripe. As illustrated in Figure 3.1, an ID-1 identification card is a rectangle approximately 85.6 mm wide, 53.98 mm tall, and 0.76 mm thick. ISO/IEC 7810 places stringent restrictions on the distortions allowed in the card backing, particularly near the area that may contain a magnetic stripe. The intent of these exacting specifications is to ensure that reader or imprinting devices into which ID-1 cards are inserted can reliably read the information off a magnetic stripe and imprint the embossed numerals without causing any damage or significant deterioration of the card.

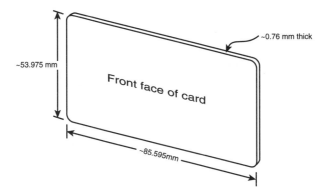

Figure 3.1. *ID-1 identification card form factor.*

The material characteristics of ID-1 identification cards are also established by ISO/IEC 7810. Specifically, the card must be composed of polyvinyl chloride, polyvinyl chloride acetate, or "materials having equal or better performance." There are correspondingly stringent deformation characteristics for the card. In general, the specifications require that after one end of a card is flexed by up to 35 mm (which corresponds to about half the width of a card), it should return to its original flat state to within 1.5 mm. Further, it is specified that it be possible to bring the card to within an acceptable flat state through the uniform application of a moderately light pressure across the face of the card. Interestingly, the actual durability of a card is not established by the ISO/IEC specifications, but rather is left to a "mutual agreement between the card issuer and the manufacturer."

Encoding of Information for Identification Cards

Building on the base formed by ISO/IEC 7810, the ANSI/ISO/IEC 7811 specification establishes standards for the encoding of information on an identification card through embossing or magnetic stripe techniques.

Note

ISO/IEC standards are available from

ISO/IEC Copyright Office
Case Postale 56
CH-1211 Geneve 20
Switzerland

Copies of international standards, catalogs, and handbooks (ISO and IEC), as well as all foreign standards from ISO member body countries (DIN, JISC, BSI, AFNOR, and so on), are available in the United States from

ANSI
11 West 42nd Street
New York, NY 10036
212-642-4900 (voice)
212-302-1286 (fax)

Embossing means causing the shape of characters to rise above the backing plane of the card. The embossed characters essentially form a typeface that can be used to print these characters onto some other material through the use of a rudimentary "printing press." Before online printing of transaction receipts became prevalent, merchants used these imprinting devices to prepare credit card invoices and receipts. ANSI/ISO/IEC 7811-1 establishes the allowed characteristics of embossing itself, including the relief height of the embossed characters (0.46 mm to 0.48 mm), the spacing between embossed characters (2.54 mm to 3.63 mm), and the size of the characters (4.32 mm). Auxiliary ISO specifications identify the characters and font sizes that render embossed characters suitable for optical recognition devices, along with the test procedures used to determine that a specific identification card meets all of these specifications. These include

- ISO 1073-1, Alphanumeric character sets for optical recognition—Part 1: Character set OCR-A—Shapes and dimensions of the printed image
- ISO 1073-2, Alphanumeric character sets for optical recognition—Part 2: Character set OCR-B—Shapes and dimensions of the printed image
- ISO 1831, Printing specifications for optical character recognition
- ISO/IEC 10373, Identification cards—Test methods

ANSI/ISO/IEC 7811-2 specifies the recording techniques used to encode characters into a magnetic stripe affixed to an ID-1 card. Provisions are made for three different types of information recording, referenced as Track 1, Track 2, and Track 3. Track 1 can contain up to 79 alphanumeric characters encoded at a write density of 8.27 bits per mm (210 bits per inch). This track can contain both alphabetic and numeric information. Track 2 can contain up to 40 characters of numeric information encoded at a write density of 2.95 bits per mm (75 bits per inch). Both Track 1 and Track 2 are intended to be write-once/read-many channels—essentially, once the card is issued, these are read-only channels. Track 3 is both a write-many and a read-many facility (that is, a read/write track). It can contain up to 107 characters encoded at 8.27 bits per mm (210 bits per inch). The encoding for each of the data tracks includes a "longitudinal redundancy check" character that can be used by the card reader to detect any errors in the information read versus what was originally written onto the card.

ANSI/ISO/IEC 7811-3 specifies in detail the location of embossed characters on an ID-1 card, and Part 4 specifies the location of magnetic stripes. As illustrated in Figure 3.2, two areas for embossing are specified. The first, whose center line is 21.42 mm above the bottom edge of the card, or just below the center line of the card, allows for up to 19 card identification number numerals to be embossed. Just below this is an additional area of approximately 14.53 mm by 66.04 mm in which 4 rows of 27 characters each can be used to form a name and address field. This is offset at least 2.41 mm from the bottom of the card and 7.65 mm from the left edge; the embossed characters are raised toward the front side of the card. If a magnetic stripe is included on the card, it is found near the top, on the back side of the card. The specifications state that the magnetic stripe and the embossing may not overlap.

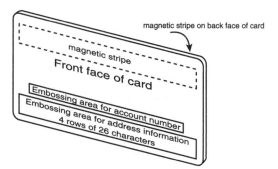

Figure 3.2. *Embossing and magnetic stripe locations.*

Two variants of magnetic stripes can be found on ID-1 identification cards; the form and location of these are defined in ANSI/ISO/IEC 7811-4 (for read-only tracks) and Part 5 (for read/write tracks). One of these is 6.35 mm tall by 79.76 mm wide, positioned no more than 5.54 mm from the top edge of the card and on the back face of the card. This magnetic stripe supports two recording tracks, each of which is intended to be a read-only track.

The Business Model for Identification Cards

By following the ISO standards through several interconnected specifications for identification cards, it is possible to go beyond just the description of physical and electronic characteristics of the card. They have arrived at a business model from which inferences can be made regarding how cards will be manufactured, what groups will actually distribute the cards to end users, and some of the operations to be performed by the end users of the identification cards. For example, the ANSI/ISO/IEC 7811-1 specification defines two terms reflecting the "distribution state" of a card:

- *Unused card*—A card that has been embossed with all the characters required for its intended purpose but has not been issued.
- *Return card*—An embossed card after it has been issued to the cardholder and returned for the purpose of testing.

ANSI/ISO/IEC 7811-2 further defines similar states for magnetic stripe cards:

- *Unused unencoded card*—A card possessing all components required for its intended purpose that has not been subjected to any personalization or testing operation. The card has been stored in a clean environment without more than 48-hour exposure to daylight at temperatures between 5 degrees C and 30 degrees C and humidity between 10% and 90%, without experiencing thermal shock.

- *Unused encoded card*—An unused, unencoded card that has only been encoded with all the data required for its intended purpose (for example, magnetic encoding, embossing, electronic encoding).

- *Returned card*—An embossed or encoded card after it has been issued to the cardholder and returned for the purpose of testing.

ANSI/ISO/IEC 7812: "Identification of Issuers—Part 1: Numbering System" further develops the business model by establishing a standard for the card identification number, which is displayed in embossed characters on the front face of an ID-1 card. The card identification number, which may be up to 19 characters long, is subdivided into three components:

- Issuer identification number—A six-digit component that includes the following:
  - Major industry identifier—A one-digit indicator of the industry designation of the card issuer; it is one of the following:

    0—Tag reserved to indicate new industry assignments

    1—Airlines

    2—Airlines and other future industry assignments

    3—Travel and entertainment

    4—Banking/financial

    5—Banking/financial

    6—Merchandizing and banking

    7—Petroleum

    8—Telecommunications and other future industry assignments

    9—For assignment by national standards bodies

- Issuer identifier—A five-digit number associated with the specific issuing organization.

- Individual account identification number—A variable-length component up to 12 digits maximum.

- Check digit—A cross-check number that is calculated from all the previous digits in the identification number according to an algorithm called the *Luhn formula*, which is defined in an appendix of ANSI/ISO/IEC 7812.

The path toward standards-based specification of a general business mode (for financial transactions) becomes very explicit with ISO/IEC 7813: Identification Cards—Financial Transaction Cards. This specification does not consider any new technical areas, but makes a strict enumeration of the standards that must be adhered to in order to call a card a *financial transaction card*.

ISO/IEC 7813 specifies the content of the two read-only tracks of a magnetic strip included on the card. This augments the content definition for ISO 4909 for the read/write track. The end result is a complete description of both the technical characteristics and the information content of cards suitable to support financial transactions, all rooted in international standards and acceptable for worldwide deployment.

Smart Card Standards

ISO 7816: Identification Cards—Integrated Circuit(s) Cards with Contacts provides the basis to transition the relatively simple identification card from a token that can be compromised through forgery, theft, or loss into a tamper-resistant and "intelligent" *integrated circuit card* (ICC), more popularly known as a *smart card*. It is a multiple-part standard through which the smart card is specified in sufficient detail to achieve the same level of interoperability that has been achieved with the simpler cards discussed in the section "ISO Standards for Cards." Although ISO 7816 includes at least six approved parts and has several additional parts under review, the discussion here is limited to Parts 1 through 5:

- Part 1—Physical characteristics
- Part 2—Dimensions and location of the contacts
- Part 3—Electronic signals and transmission protocols

- Part 3, Amendment 2—Revision of protocol type selection
- Part 4—Inter-industry commands for interchange
- Part 5—Numbering system and registration procedure for application identifiers

Note

In the ISO standards related to ICCs, the device into which the ICC is inserted is referred to as an *interface device* (IFD). In the course of this book, the terms *ICC*, *card*, and *smart card* tend to be used interchangeably. Similarly, the terms *IFD*, *reader*, and *terminal* are used to mean the same thing.

ISO 7816-1 extends the physical characteristics definition of simpler ID-1 identification cards from the realm of plastic cards with perhaps an associated magnetic stripe to the more complex environment supporting an integrated circuit chip within the card. This includes accommodation of exposure limits for a number of electromagnetic phenomena such as x-rays, ultraviolet light, electromagnetic fields, static electrical fields, and ambient temperature of the card (with embedded chip), as indicated in Table 3.2.

Table 3.2. Exposure limits for physical phenomena.

Phenomenon	Limit
Ultraviolet light	Ambient (depends on card vendor)
X-rays	Two times acceptable annual human dosage
EMI (electromagnetic interface)	No interference with magnetic stripe
Electromagnetic fields	Less than 1,000 Oe
Static electricity	1,500 volt discharge through 1.5 K ohm resistor from 100 pF capacitor
Heat dissipation	Less than 2.5 watt; card temperature less than 50° C

The specification is also concerned with defining the characteristics of the card when it is bent or flexed. The concern, of course, is that an environment amenable to operation of the chip in the card—with its micro-electronic connections between surface connectors and chip I/O pins, as well as its integrated circuitry—be maintained. Two tests of flexibility for the card are specified. Figure 3.3

illustrates the bending requirements of the card and Figure 3.4 illustrates the torsion requirements of the card.

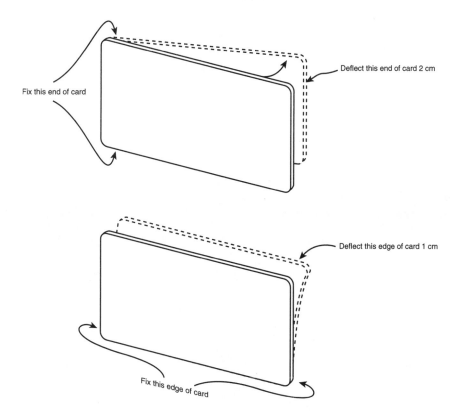

Deflect this end of card 2 cm

Fix this end of card

Deflect this edge of card 1 cm

Fix this edge of card

Figure 3.3. *Card bending testing.*

In the cases of both bending tests and torsion tests, the concern is that through the normal wear and tear on a card (for example, keeping the card in one's wallet), either the chip itself or the microconnection wires from the chip to the surface contacts will be damaged or broken. Practical experience with these tests has shown that a chip size on the order of 25 mm^2 is the largest that can routinely meet these flexibility constraints.

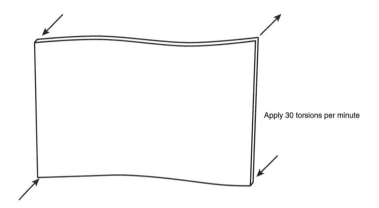

Apply 30 torsions per minute

Figure 3.4. *Torsion testing of a smart card.*

Characteristics of Smart Cards

The mechanical tolerances required of smart cards are relatively stringent in order to ensure that cards can be properly aligned so a reader's contact points can make a good electrical connection with the correct card contacts. In general, the card's contacts cannot vary from the surface of the card by more that 0.1 mm. The card must be sufficiently strong to resist permanent deformation when bent; it must be possible to return the card to a flat position with a very modest pressure over the face of the card. Finally, the electrical resistance of the card's contacts must fall within the acceptable limits established by ISO 7816-1.

ISO 7816-2 specifies an ICC with eight electrical contacts present in a standardized position on the front face of the card. These are referred to as C1 through C8. Some of these contacts are electrically connected to the microprocessor chip embedded within the card; some are not, having been defined to allow for enhancements but unused at the present time. The specific definitions for the contacts are shown in Table 3.3.

Table 3.3. Contact definitions for smart cards.

Contact	Designation	Use
C1	V_{CC}	Power connection through which operating power is supplied to the microprocessor chip in the card.
C2	RST	Reset line through which the IFD can signal to the smart card's microprocessor chip to initiate its reset sequence of instructions.

Contact	Designation	Use
C3	CLK	Clock signal line through which a clock signal can be provided to the smart card's microprocessor chip to control the speed at which it operates and to provide a common framework for data communication between the reader and the smart card.
C4	RFU	Reserved for future use.
C5	GND	Ground line providing a common electrical ground between the reader and the smart card.
C6	V_{PP}	Programming power connection providing a separate source of electrical power (from the operating power) that can be used to program the nonvolatile memory on the microprocessor chip.
C7	I/O	Input/output line that provides a half-duplex communication channel between the reader and the smart card.
C8	RFU	Reserved for future use.

The location of the contacts is illustrated in Figure 3.5. The contacts are almost always found on the front face of the card, which is the side of the card with the primary graphic and opposite the side with any magnetic stripe. However, the ISO 7816-2 standard does not mandate that the contacts appear on the front; the contacts can appear on the back of the card as long as care is taken to make sure the contacts do not intersect the magnetic stripe area.

Many of the earliest smart cards adhered to a different standard for contact locations that positioned the contacts toward the upper-left portion of the front face of the card. The standard on which this positioning was based became obsolete in 1990. Cards designed according to this standard were deployed primarily in Europe to support credit and debit applications.

ISO 7816-3 begins to delve into the specification of the "intelligent" aspects of the smart card. This standard describes the relationship between the smart card and the reader as one of "slave" (the smart card) and "master" (the reader). Communications are established by the reader signaling to the smart card through the contacts noted previously and are continued by the smart card responding accordingly. Communication between the card and reader proceed according to various state transitions illustrated in Figure 3.6. The communication channel is single-threaded; once the reader sends a command to the smart card, it blocks until a response is received.

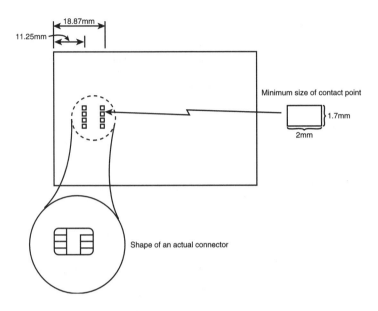

Figure 3.5. *Location, size, and shape of contacts.*

When a card is inserted into a reader, no power is applied to any of the contacts. The chip on the card could be seriously damaged by applying power to the wrong contacts, and this situation could easily occur if a card were inserted across powered contact points. The contacts remain unpowered until an edge detector determines that the card is properly aligned with the contact points to within some acceptable (for the reader) mechanical tolerance.

Note

Security mechanisms on many chips could be triggered if the cards were inserted across powered contact points, resulting in the possible disabling of the chip.

When the reader detects that the card is properly inserted, power is applied to the card. First, the contacts are brought to a coherent idle state, as shown in Table 3.4. A reset signal is then sent to the card via the RST contact line. The idle state is characterized as being when the power (V_{CC}) contact is brought up to a normal, stable operating voltage of 5 v. An initial power setting of 5 v is always applied first, even though some microprocessor chips being introduced operate at 3 v when in an I/O state. The I/O contact is set to a reception mode on the reader side and a stable clock (CLK) is applied. The reset line is in a low state. It must remain in a low state for at least 40,000 CLK cycles before a valid reset sequence can be started by the reader, raising the reset line to a high state.

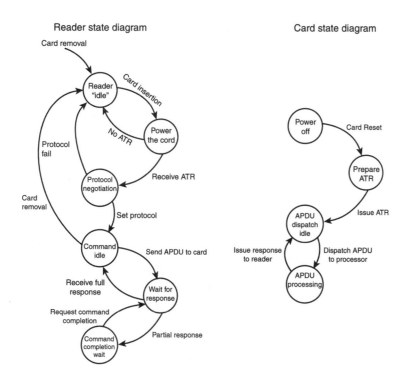

Figure 3.6. *Reader and smart card general state diagrams.*

Table 3.4. Contact states prior to card reset.

Contact	State
V_{CC}	Powered and stable
V_{PP}	Stable at idle state
RST	State—low
CLK	Suitable and stable clock signal applied
I/O	Reception mode in interface device

As illustrated in Figure 3.7, powering up a smart card occurs according to a well-defined sequence. Once power has been satisfactorily applied to the card, the RST line is raised to a high state that signals to the card to begin its initialization sequence. The specific initialization operations can vary from card to card, but the sequence should result in the sending of an answer to reset (ATR) from the card back to the reader. In general, the first byte of the ATR must be received by the reader within 40,000 clock cycles. Following that, each successive byte of the ATR must be received by the reader at a rate of at least 1 byte per second.

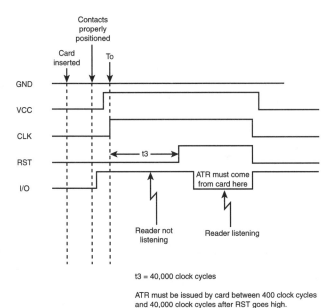

Figure 3.7. *The reader reset sequence.*

Data transfer between the reader and the card occurs through the concerted action of two of the contact lines: CLK and I/O. The I/O line conveys a single bit of information per unit of time as defined by the CLK depending on its voltage relative to GND. A 1 bit can be conveyed either through a +5 v value or through a 0 v value. The actual convention used is determined by the card and is conveyed to the reader through the "initial character" of the ATR, which is referenced as *TS*. To transfer 1 byte of information, 10 bits are actually moved across the I/O line; the first is always a "start bit" and the last is always a parity bit used to convey even parity. Considering that the I/O line can be (in one bit period) either in a high (H) state or a low (L) state, the TS character of the form HLHHLLLLLH signals that the card wants to use the "inverse convention," meaning that H corresponds to a 0 and L corresponds to a 1. A TS character of the form HLHHLHHHLLH signals that the card wants to use the "direct convention," meaning that H corresponds to a 1 and L corresponds to a 0.

The direct convention and the inverse convention also control the bit ordering with each byte transferred between the card and the reader. In the direct convention, the first bit following the start bit is the low-order bit of the byte. Successively higher-order bits follow in sequence. In the inverse convention, the first bit following the start bit is the high-order bit of the byte. Successively lower-order bits follow in sequence. Parity for each byte transferred should be even; this means that the total number of 1 bits in the byte, including the parity bit, must be an even number.

The I/O line comprises a half-duplex channel; that is, either the card or the reader can transmit data over the same channel, but they both cannot be transmitting at the same time. So as part of the power-up sequence, both the reader and the card enter a receive state in which they're listening to the I/O line. With the commencement of the reset operation, the reader remains in the receive state while the card must enter a send state in order to send the ATR back to the reader. From this point on, the two ends of the channel alternate between send states and receive states. With a half-duplex channel, there is no reliable way for either end to asynchronously change a state from send to receive or from receive to send. Rather, if this is desired, that end must go into a receive state and allow a time-out of the operation in progress; then a reader end will always try to re-establish a known sequence by re-entering a send state.

The CLK and I/O lines can support a wide variety of data transmission speeds. The specific speed is defined by the card and is conveyed back to the reader through an optional character in the ATR. The transmission speed is set through the establishment of *one bit time* on the I/O line, which means that an interval is established at which the I/O line may be sampled in order to read a bit and then each successive bit. This time is defined as an *elementary time unit* (etu) and is established through a linear relationship between several factors. Note that the TS character was returned before any definition of the etu could be made. This is possible because the etu during the ATR sequence is always specified to be

$$etu_0 = 372/(\text{CLK frequency})$$

where the CLK frequency is always between 1 MHz and 5 MHz; in fact, the frequency is almost always selected such that the initial data transfer rate is 9,600 bits per second.

Once an RST signal is sent from the reader to the card, the card must respond with the first character of the ATR within 40,000 CLK cycles. The card might not respond with an ATR for a number of reasons, the most prevalent being that the card is inserted incorrectly into the reader (probably upside down). In some instances, the card may not be functioning because it has been damaged or broken. Whatever the case, if the ATR is not returned within the prescribed time, the reader should begin a sequence to power down the card. During this sequence, the reader sets the RST, CLK, and I/O lines low and drops voltage on the V_{CC} line to nominal 0 (that is, less than 0.4 v).

The ATR is a string of characters returned from the card to the reader following the successful completion of the power-up sequence. As defined in ISO/IEC 7816-3, the ATR consists of 33 or fewer characters comprising the following elements:

- TS—a mandatory initial character
- T0—a mandatory format character
- TA_i TB_i TC_i TD_i—optional interface characters
- T1, T2,...TK—optional historical characters
- TCK—a conditional check character

The historical characters can be defined at the discretion of the card manufacturer or the card issuer. These characters are typically used to convey some type of designation of the type, model, and use of this specific card. When used in this way, the historical characters provide a modest mechanism through which systems can automatically detect the use of an inserted card (within that system) and can initiate other actions (or software) accordingly. The check character provides a mechanism through which the integrity of the ATR can be measured; that is, whether a transmission error has occurred in sending the characters from the card to the reader.

The structure of the ATR is illustrated in Table 3.5. As discussed previously, the initial TS character is used to establish the bit-signaling and bit-ordering conventions between the reader and the card. The T0 character is used to signal the presence or absence of subsequent interface characters or historical characters. The interface characters are used to tailor the characteristics of the I/O channel, including the specific protocol used by the card and reader during subsequent exchange of commands (from the reader to the card) and responses (from the card to the reader). The historical characters, if present, are used to convey card-manufacturer–specific information from the card to the read, and hence to the application system being served by the reader. There is really no established standard for the information presented in the historical bits.

Table 3.5. The answer-to-reset structure.

Character ID	Definition
Initial Character Section	
TS	Mandatory initial character
Format Character Section	
T0	Indicator for presence of interface characters
Interface Characters Section	
TA_1	Global, codes F1 and D1
TB_1	Global, codes 11 and PI1
TC_1	Global, code N
TD_1	Codes Y_2 and T
TA_2	Specific
TB_2	Global, code PI2
TC_2	Specific
TD_2	Codes Y_3 and T
TA_3	TA_i, TB_i, and TC_i are specific
$...TD_i$	Codes Y_{i+1} and T
Historical Character Section	
T1	(Maximum of 15 characters)
...TK	Card-specific information
Check Character Section	
TCK	Optional check character

The interface characters are used as a selection method for the protocol used for subsequent higher-level communication between the reader and the card. Two such protocols are defined by ISO 7816-3—the T=0 protocol and the T=1 protocol. T=0 is an asynchronous protocol, meaning there is no strict timing connection between one command sent from the reader to the card and the next command sent from the reader to the card. When the card receives a command from the reader, it performs the requested operations and sends back to the reader a response relative to that command. The reader is then free to send the next command to the card whenever it needs to. The T=1 protocol is an asynchronous block transmission protocol. This means that in one transmission packet from the reader to the card, one to several commands are sent. The card responds to this (these) commands, at which point the reader can send another command or block of commands. Designations for additional protocols are defined as indicated in Table 3.6.

Table 3.6. Protocol designations.

Designation	Definition
T=0	Asynchronous (single) command/response protocol
T=1	Asynchronous (multiple) command/response protocol
T=2	Reserved for future full-duplex protocol
T=3	Reserved for future full-duplex protocol
T=4	Reserved for enhanced asynchronous protocol
T=5	Reserved for future use
T=6	Reserved for future use
T=7	Reserved for future use
T=8	Reserved for future use
T=9	Reserved for future use
T=10	Reserved for future use
T=11	Reserved for future use
T=12	Reserved for future use
T=13	Reserved for future use
T=14	Reserved for vendor-defined protocol
T=15	Reserved for future extension

The ATR sequence that initializes the physical communication channel between the reader and the card allows a number of characteristics of the channel to be defined or manipulated. The ISO/IEC 7816-3 standard defines a more elaborate adjunct to the ATR sequence called the *Protocol Type Selection* (PTS) facility. Actually, the PTS can be thought of as an extension of the ATR. Through the PTS the reader side of the channel and the card side of the channel can negotiate to an optimum set of characteristics for the channel. For most current smart card systems, there is a strong correlation between the reader-side development and the card development. Consequently, the optimum communication characteristics are almost always derived through the ATR sequence without performing a PTS sequence. Therefore, we will not delve into the details of the PTS. However, in the future, with an expanding marketplace for smart cards encouraging more disconnected development of readers, terminals, and cards, the need to go through a PTS sequence on card initialization may significantly increase.

Other Smart Card Standards and Specifications

ISO 7816 is unquestionably the most widely known and followed general-purpose smart card standard, but it is by no means the only one. There are standards for the use of smart cards in specific applications such as health, transportation, banking, electronic commerce, and identity. And there are standards for new kinds of smart cards such as proximity and contactless smart cards. Since a smart card is always part of a larger information technology, it is subject to a wide range of information-processing standards such as character sets, country encodings, monetary representations, and cryptography. Finally, since many smart card applications intersect a number of governmental concerns such as monetary systems, national identity, and benefit eligibility, there are national and regional smart card standards in addition to international standards.

Due to the deliberate pace of international standards efforts, there are also a growing number of smart card specifications issued by organizations such as governmental laboratories, professional societies, trade associations, academic institutions and private firms not associated with standards bodies. These specifications have no force other than the force of the marketplace of products and ideas, but they do serve the useful role of stimulating discussion and consensus, which can be fed into official standards efforts. As smart cards are embraced within other technologies such as cellular telephones, watches, automobiles, and Internet browsers, we can expect that these technologies and their application domains will prompt rules and regulations about the nature of the smart card component. For the near term as smart card usage explodes, it will probably be as much the marketplace as the international standards process that will say how smart cards are supposed to be.

As by no means an exhaustive list but rather to simply give you some starting points for further search, Tables 3.7 and 3.8 list some further standards and specifications that are influencing the deployment and evolution of smart cards.

Table 3.7. Examples of smart card standards.

Standard	Subject Area
International Standards Organization	
ISO 639	Languages, countries, and authorities
ISO 646	7-bit coded character set
ISO 3166	Names of countries
ISO 4217	Currencies and funds

continues

Table 3.7. continued

Standard	Subject Area
International Standards Organization	
ISO/IEC 7501	Travel documents
ISO/IEC 7810,7811, 7812	Magnetic stripe cards
ISO/IEC 7813	Financial transactions
ISO 8601	Dates and times
ISO 8859	8-bit coded character set
ISO 9564	PIN management
ISO 9797	Data cryptographic techniques
ISO 9992	Messages between card and terminal
ISO 10202	Financial transaction cards
ISO 10536	Contactless integrated circuit cards
ISO 11568	Cryptographic key management
ISO 11694	Optical memory cards
European Telecommunications Standards Institute (ETSI)	
ETSI TE9	Card terminals
GSM 11.4	Subscriber identification module (SIM) cards for GSM cellular telephones
European Committee for Standardization (CEN)	
TC 224	Machine-readable cards
EN 726	Requirements for IC cards and terminals for telecommunications use
Commission of the European Union (CEU)	
ITSEC	Information technology security evaluation criteria
European Computer Manufacturers Association (ECMA)	
ECMA-219	Key distribution
International Telecommunication Union (ITU)	
X.400	Secure email
X.509	Authentication framework
American National Standards Institute (ANSI)	
ANSI X9.15-1990 (R1996)	Specification for financial message exchange between card acceptor and acquirer

Standard	Subject Area
American National Standards Institute (ANSI)	
ANSI X9.8-1995	Banking—personal identification number management and security, Part 1: PIN protection principles and techniques; and Part 2: approved algorithms for PIN encipherment
ANSI X3.15-1975 (R1996)	Bit sequencing of the American National Standard Code for Information Interchange in serial-by-bit data transmission
ANSI X3.118 (1984)	PIN pad specification
U.S. National Institute for Standards and Testing (NIST)	
FIPS 140-1	Cryptographic tokens

Table 3.8. Examples of smart card specifications.

Specification	Sponsor(s)	URL
Integration of Smart Cards into the Pluggable Authentication Module (PAM)	Open Group	www.opengroup.org
RFC 86.0—Unified Login with PAM	Open Group	www.opengroup.org
RFC 57.0 Smart Card Introduction	Open Group	www.opengroup.org
ISI-3 (IBM Smartcard Identification)	IBM and the University of Twente	www.iscit.surfnet.nl
International Chip Electronic Commercial Standard	Visa	www.visa.com
Integrated Circuit Card (ICC) Specification	Visa	www.visa.com
Java Card 2.0	Sun Microsystems	www.javasoft.com
IATA 791—Airline Ticketing	International Airline Travel Association	www.iata.org
PKCS #11 Cryptographic Token Interface Standard	RSADSI	www.rsa.com
Electronic ID Application	Secured Electronic Information in Society (SEIS)	www.seis.se

Specification	Sponsor(s)	URL
ICC Specification for Payment Systems (EMV'96)	Europay, MasterCard, Visa	www.mastercard.com
Standards 30 and 40—Card Terminals	Association for Payment Clearing Services (APACS)	
OpenCard Framework	IBM, Netscape, NCI, and Sun Microsystems	www.opencard.com
PC/SC Workgroup	Bull, HP, Microsoft, Schlumberger, and Siemens Nixdorf	www.smartcardsys.com

Summary

This chapter describes the international standards basis for the characteristics and form factors of identity cards, including a variety of mechanisms for encoding information used in transactions that the cards must support. This chapter also considers the addition of integrated circuit chips to such cards, the mechanisms involved in establishing a communication channel between a reader of such cards, and the cards themselves. Chapter 4, "Smart Card Commands," details the two most prevalent protocols used between readers and cards (the T=0 protocol and the T=1 protocol).

CHAPTER 4

SMART CARD COMMANDS

Chapter 3, "Some Basic Standards for Smart Cards," examines the foundation for smart card interoperability found in a variety of international standards. Through the answer-to-reset (ATR) mechanism, a basic communication pathway is established between a reader and a smart card. This communication pathway is a half-duplex physical channel on which either the reader can talk and the card listen, or the card can talk and the reader listen. This chapter examines the establishment and use of more complex protocols on top of this physical channel.

Layered directly on top of the physical channel is a *link-level* communication protocol that provides an error-free connection through which the reader can initiate actions on the card by issuing a command to the card. The card can then perform some sequence of operations, as dictated by the command from the reader, and can send a response back to the reader as well as provide a status indication regarding the command. While the standards allow for the specification of 15 or more distinct link-level protocols, there are actually two primary variants of this higher level protocol in use with smart card systems today: the T=0 protocol and the T=1 protocol.

Once the link-level protocol (T=0, T=1, and so on) is established, application-level protocols—that make use of the link level to communicate between application elements on the card and other application elements on the reader—can be defined. The ISO/IEC 7816-4 standard defines two such application protocols: One is aimed at providing a consistent file system for the storage and retrieval of information on a card. The application programming interface (API) that constitutes this protocol encompasses a set of function calls through which the file system can be manipulated; files can be selected by name, information can be written to files, information can be read from files, and so on. A second protocol in the form of an API for accessing security services on a card is also defined by ISO/IEC 7816-4. This API allows the card and the reader to mutually authenticate each other's identity and provides them with mechanisms through which the two sides can keep private the information that they exchange.

To support the application protocol APIs, a protocol message structure is defined in ISO/IEC 7816-4, through which the function calls, their associated parameters, and status response parameters are exchanged between the reader-side application and the card-side application code. This message structure is characterized by application protocol data units (APDUs), which are conveyed between the reader- and card-side application code by the link-level protocol (generally either a T=0 or a T=1 protocol).

While the qualitative semantics of the file access and security APIs will be described in this chapter, a more detailed description of these APIs can be found in Appendix A, "The ISO/IEC 7816-4 Command Set."

Link-Level Protocols

When talking about communications protocols, you can generally analyze the situation in terms of the OSI (Open Systems Interconnect) Reference Model, which is shown in Figure 4.1. The OSI Reference Model describes the general communication problem between two entities in terms of seven distinct protocols that layer on top of each other (hence the term *seven-layer model*) and provide a complete mechanism through which two applications on disparate platforms can effectively exchange information with each other. A central theme of the OSI Reference Model is the strict separation of layers. That is, a layer communicates only with the layer immediately above or below it through a well-defined interface, and each layer provides a specific set of services to the entire protocol stack. In the case of the T=0 and T=1 smart card protocols, the T=1 protocol fits the OSI Reference Model fairly well as a "data link" (or link-level) protocol layer, but the T=0 protocol tends to mix elements from several different protocol layers (as defined by the OSI Reference Model).

Figure 4.1. *The OSI Reference Model.*

At this point, it should be reiterated that the reader and the card communicate through a very strict command-and-response protocol. That is, the reader-side of the link sends a command to the card, possibly including data to be used by the card in the execution of the command, and the card then executes that command and sends a response back to the reader. This response may include data resulting from the execution of the command on the card as well as a status response regarding the execution of the command. The data structures exchanged by the reader and the card in this command-and-response protocol are referred to as *transmission protocol data units* (TPDUs). The TPDU structures used in the T=0 protocol are quite different from those used in the T=1 protocol.

Once the T=0 or T=1 protocol is established between the reader and the card, it is used to support application-level protocols between application software on the card and application software on the reader-side of the link. These application protocols exchange information through data structures referred to as *application protocol data units* (APDUs). The details of the various TPDU and APDU structures are reviewed in the following sections. One conclusion that can be drawn from this discussion is the fact that the T=0 protocol provides very poor layer separation between the link-level protocol and the application-level protocol. Consequently, it is found that mechanisms that would normally be based on intervening protocol layers (between the application layer and the data link layer) can be extremely awkward to implement on top of the T=0 protocol. Principal among these is "secure messaging" between application software on the card with application software on the reader-side of the link. *Secure messaging* refers to the use of cryptographic techniques to limit access to information conveyed between application components on the reader-side and application elements on the card side of the channel. Specifically, only the application components on each end of the channel should be able to understand the information transferred between them; the intervening layers should simply see the information flowing as unintelligible collections of bytes.

The poor protocol layering of the T=0 protocol is not the product of poor design. Rather, it is the result of an attempt to make the protocol as responsive as possible in order for communication between the reader and the card to be as efficient as possible. The data transmission speed across the reader-to-card interface is relatively slow (nominally 9600 bits per second) and this channel is in the critical path of all transactions that involve the card. To maximize consumer satisfaction, it is desirable that such transactions proceed as quickly as possible—instantaneously would be very nice. Consequently, in the T=0 protocol, the error handling and the application protocol support are optimized to minimize the amount of information which flows across the reader-to-card interface and thereby minimize the transaction time.

The T=0 Protocol

The T=0 protocol is a byte-oriented protocol. Like all other ISO-compliant smart card protocols, it functions in a command-response mode in which the reader-side of the connection issues a command to the card, which then performs the commanded operation and sends back a response.

Note

Byte-oriented means that a byte is the unit of information transferred across a channel and that error handling is handled one byte at a time as well.

In the T=0 protocol, error detection is done by looking at the (even) parity bit on each byte transferred across the reader-to-card interface. The transfer of each byte of information requires the use of 10 bits, as illustrated in Figure 4.2. The parity bit is cleared or set to make the total number of bits set (per character transferred) be an even number. The receiver side of the channel can look at the bit values transferred prior to the parity bit and determine whether the parity bit should be set. If the actual parity bit transferred does not match what was expected, then it can be assumed that an error exists in the byte of data just transferred and some recovery procedure must be undertaken. The recovery procedure used with the T=0 protocol is triggered by the receiving side, which, on detecting a parity error, signals that it expects the transmitting side to retransmit the byte (that was received in error). It provides the signal to the transmitting side by holding the I/O line in a low state. Normally, the I/O line is in a high state immediately preceding the transfer of a byte, so a low state acts as an error feedback signal to the transmitter. On detecting this, the transmitting side of the channel waits for at least two character times and then again sends the byte that was previously received in error.

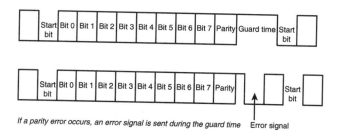

If a parity error occurs, an error signal is sent during the guard time   Error signal

Figure 4.2. *Reader/card byte transfer and error feedback loop.*

Those well versed in communication protocols might see this error detection and recovery mechanism as being somewhat prone to less-than-perfect behavior. This indeed tends to be the case in actual practice. For most readers, however, the channel tends to be very good or very bad. If it's very good, then this error detection and recovery mechanism is seldom used; if it's very bad, then the error detection and recovery mechanism is likely to fail at some point. This leads to the transmitting and receiving sides of the channel getting out of synchronization. If this situation is detected by the card, it will usually be programmed to go mute and quit responding to commands from the reader. At this point, or if the reader detects the ambiguous state first, the reader issues a reset signal to the card that forces the communication protocol to be brought up from scratch.

The TPDU for the T=0 protocol comprises two distinct data structures: one that is sent from the reader to the card (as a command) and one that is sent from the card to the reader (as a response). The command header (sent from the reader to the card) includes five fields:

- CLA —A 1-byte field that establishes a collection of instructions; this is sometimes referred to as the *class designation* of the command set.

- INS—A 1-byte field that specifies a specific instruction (to the card) from within the set of instructions defined within the CLA designation; this is sometimes referred to as the *instruction designation* within the class of commands.

- P1—A 1-byte field used to specify the addressing used by the [CLA,INS] instruction.

- P2—A 1-byte field also used to specify the addressing used by the [CLA,INS] instruction.

- P3—A 1-byte field used to specify the number of data bytes transferred either to the card or from the card as part of the [CLA,INS] instruction execution.

The *procedure bytes,* which are sent from the card to the reader, are used to respond to the reader's command and include three required fields and one optional field:

- ACK—A 1-byte field that indicates reception (by the card) of the [CLA,INS] command.
- NULL—A 1-byte field used by the card to essentially do flow control on the I/O channel; it sends the message from the card to the reader that the card is still working on the command and signals the reader not to send another command just yet.
- SW1—A 1-byte field used by the card to send a status response back to the reader regarding the current command.
- SW2—A 1-byte (optional) field that may be included in the procedure bytes, depending on the specific command being executed. If included, it also conveys a status response back to the reader.

As indicated previously, the T=0 protocol tends to mix elements of application-level protocols with elements of link-level protocols. The definition of the CLA byte is one such case in point. Each value of CLA defines an application-specific set of instructions; the individual instructions have a unique INS value. The first set of application-oriented commands is found in ISO/IEC 7816-4; the specific command sets are aimed at manipulation of a file system on a card and at accessing "security" commands on a card. Other standards documents define additional sets of commands (that is, additional values of the CLA byte). Some of these are listed in Table 4.1. The specific instructions found in these classes will be reviewed later in this chapter.

Table 4.1. CLA instruction set definitions.

CLA Byte	Instruction Set
0X	ISO/IEC 7816-4 instructions (files and security)
10 to 7F	Reserved for future use
8X or 9X	ISO/IEC 7816-4 instructions
AX	Application- and/or vendor-specific instructions
B0 to CF	ISO/IEC 7816-4 instructions
D0 to FE	Application- and/or vendor-specific instructions
FF	Reserved for protocol type selection

Within a given CLA value (that is, within a class of instructions identified by a common value of CLA), the INS byte is used to identify a specific instruction. As indicated in Table 4.1, several different standards identify collections of instructions. The ISO/IEC 7816-4 standard identifies a number of instructions used to access an on-card file system and security functions that serve to limit access to the file system and to the card in general. This instruction set is listed in Table 4.2.

Note

At this point, a portion of the discussion from Chapter 2, "Physical Characteristics of Smart Cards," related to how EEPROM memory is programmed should be reiterated. In early microprocessor chips that included EEPROM memory, a separate power source (V_{pp}) was needed to program (that is, erase and write) EEPROM memory. Integrated circuit chips used in current smart cards are able to derive "programming power" for nonvolatile memory from the V_{cc} power, so any significant discussion regarding V_{pp} has been omitted. When the ISO/IEC 7816 standard was adopted, however, the proper manipulation of the V_{pp} power was necessary and was subsequently embedded rather deeply into the T=0 protocol. Specifically, the manner in which the INS byte is defined and in which the ACK procedure byte is returned are, in effect, the control mechanisms for V_{pp}. Suffice it to say that all INS values must be even because the low order bit allows control over V_{pp} to be exercised. Further, in the absence of any manipulation of V_{pp}, the ACK procedure byte is always returned as an exact copy of the INS byte in the command TPDU to which the procedure bytes form a response.

Table 4.2. ISO/IEC 7816-4 INS codes.

INS Value	Command Name
0E	Erase Binary
20	Verify
70	Manage Channel
82	External Authenticate
84	Get Challenge
88	Internal Authenticate
A4	Select File
B0	Read Binary
B2	Read Record(s)
C0	Get Response
C2	Envelope
CA	Get Data
D0	Write Binary
D2	Write Record
D6	Update Binary
DA	Put Data
DC	Update Record
E2	Append Record

There are additional constraints on the values the INS byte can take; specifically, the high-order half-byte cannot have the value of either 6 or 9. In both cases, the restricted values are related to control mechanisms used to manipulate the V_{PP} power source. See the ISO/IEC 7816-4 standard for further information.

The command header parameters P1 and P2, although defined at what should be the link-protocol level, are actually dependent (for their specific definition) on the actual instruction specified; that is, their definition is actually dependent on application protocol information. P1 and P2 provide control or addressing parameters for the various application-specific instructions. For example, one application instruction (which is examined later in this chapter) involves the selection of a specific file within the card's file system; selecting a file allows subsequent operations, such as reading or writing, to be performed on the selected file. For this specific instruction, the parameter P1 is used to control how the file is referred to in the select operation (that is, whether it will be referred to by an identifier, by name, or by path). When the Select File instruction is reviewed, the strict definitions of those terms will be considered; however, readers familiar with general file systems on various computer systems can readily infer the meanings. For the Select File instruction, the parameter P2 offers further refinement of which file is to be selected.

The command header parameter P3 is also an application-level parameter. For many instructions, the P3 parameter can take on rather complex connotations (for example, multiple parameters are defined within it). When the TPDU structure is examined, it is found that P3 generally defines the number of data bytes that are to be transmitted during the execution of the INS specified instruction. The direction of movement of these bytes is dependent on the instruction. The convention of movement of data is card-centric; that is, *outgoing* means data moving from the card to the reader, whereas *incoming* means data moving from the reader to the card. A value of P3=0 for an instruction specifying an outgoing data transfer means that 256 bytes of data will be transferred from the card to the reader.

Each time a command TPDU is sent from the reader to the card, a response TPDU is returned from the card to the reader. This response TPDU is made up of a number of procedure bytes. The first byte of this TPDU is an ACK byte. This byte is a repeat of the INS byte from the command TPDU to which this response is made. The second byte is the NULL byte. This byte is a way for the card to mark time while it processes the indicated command. While it is processing, the reader-side of the channel is waiting for the response TPDU. If the response does not arrive within a specified time-out period, the reader may start an RST sequence to reinitialize the protocol between the reader and the card. This is prevented if at least one NULL byte is received by the reader from the card.

SW1 is a status byte from the card to tell the reader the result of the requested instruction. The allowed values for SW1 are actually defined as part of the application protocol. For certain instructions, the card may have data bytes to be returned to the reader. In this case, a second status byte labeled SW2 is returned to the reader. This acts as a trigger for the reader to now execute another command called a GetResponse command, which will actually return the data bytes generated by execution of the previous command.

As you can see, the T=0 protocol is a relatively optimized protocol for moving commands and responses between the card and reader. It tends to blur the distinctions between the application-layer protocol and the link-layer protocol, with many of its constituent elements actually being defined within the application-layer protocol.

The T=1 Protocol

The T=1 protocol is a block-oriented protocol. This means that a well-defined collection of information—or *block*—is moved as a single unit between the reader and the card. Embedded within this block structure may be an APDU defined for a specific application. This facility is a good illustration that the T=1 protocol provides excellent layering between the link protocol layer and the application protocol layer. Moving information in a block, however, requires that the block be transferred (between the reader and the card) error free, or else the protocol can easily get lost. The error detection and correction, then, is a significantly more complex operation than was the case with the T=0 protocol.

Error detection in the T=1 protocol is done by using either a *longitudinal redundancy character*, which is essentially a slightly more complex form of parity checking than was done in the T=0 protocol, or by using a *cyclic redundancy check* character, which is guaranteed to detect any single-bit errors in a transmitted block. The specific CRC algorithm used is defined in detail in the ISO 3309 standard. When an error is detected within a block by the received end of the channel, it signals the transmitting end to repeat sending the block received in error.

The T=1 protocol makes use of three different types of blocks, as illustrated in Figure 4.3. Each has the same structure, but serves a different purpose:

Prologue Field			Information Field	Epilogue Field
Node Address	Protocol Control Byte	Length	APDU	Error Detection
NAD	PCB	LEN	Data Length	LRC/CRC
1 Byte	1 Byte	1 Byte	0 to 254 Bytes	1 or 2 Bytes

Figure 4.3. *T=1 protocol components.*

- Information block—This block is used to convey information between application software in the card and application software on the reader-side of the channel.

- Receive ready block—This block is used to convey either positive or negative acknowledgments from one end of the channel to the other. A positive acknowledgment indicates that a block was correctly received while a negative acknowledgment indicates that an error was detected (via checking the LRC or CRC) in the received block.

- Supervisory block—This block is used to convey control information between the card and the reader.

Each T=1 block comprises three fields:

- Prologue field—A mandatory field in the block which is 3 bytes in length. It includes the following three elements:
  - NAD—Node address
  - PCB—Protocol control byte
  - LEN—Length

- Information field—An optional field in the block that may be up to 254 bytes in length.

- Epilogue field—A mandatory field in the block that is either 1 or 2 bytes in length.

The NAD element is used to identify the addresses of the source of the block and the intended destination for the block. This addressing facility is of greatest use when the T=1 protocol is being used to support multiple logical connections between the card and multiple application connection points on the reader-side of the channel. When used, the NAD contains two subfields:

- SAD—Source address is indicated by the low order three bits of the NAD byte.
- DAD—Destination address is indicated by bits five through seven of the NAD byte.

In situations where multiple logical channels are not being used, the NAD field is set to all zeros. The two other bits of the NAD byte, those not used for the SAD or the DAD, are used to convey information related to controlling the V_{PP} (EEP-ROM programming power).

The PCB element is used to indicate the type of block (either an information, a receive ready, or a supervisory block). The 2 high-order (most significant) bits of the PCB byte are used to denote the various types:

- A high-order bit set to 0 indicates an information block.
- The 2 high-order bits set to 1 indicates a supervisory block.
- The high-order bit set to 1 and the next bit set to 0 indicates a receive-ready block.

T=1 is a relatively complex protocol. For purposes of this discussion, the protocol is viewed as was the T=0 protocol; that is, as a reliable channel for moving APDUs between application software elements on the card and on the reader-side of the communication channel. The other protocols that can be defined through the ATR or PTS sequence (that is, T=2, ..., T=14, T=15) serve this same purpose.

Application-Level Protocols

The ISO/IEC 7816-4 standard moves from the realm of defining base system functionality for smart cards into the realm of functionality directly useful to application software found on the smart card. Two areas of functionality are addressed:

- First, a file system is defined with a completely specified hierarchical structure. A set of functions are defined; these functions comprise an API through which application software on the reader-side of the channel can access the files and information in those files within this file system.

- Second, a series of security functions are defined which can be used to limit access to application software on the card or to files and information in those files within the card's file system.

This application software makes use of a protocol to exchange control and information between the reader and the card. This protocol is based on a block structure called an APDU. These APDUs are exchanged by making use of the T=0 and T=1 link-layer protocols. A software component on the card interprets these APDUs and performs the specified operation; this architecture is illustrated in Figure 4.4.

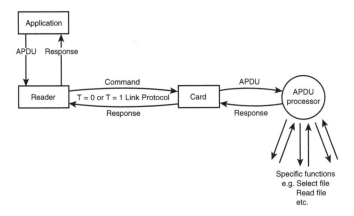

Figure 4.4. *Application communications architecture.*

The APDU structure defined in ISO 7816-4 is very similar to the TPDU structure defined in ISO 7816-3 for the T=0 protocol. When the APDU structure is transported by the T=0 protocol, the elements of the APDU directly overlay the elements of the TPDU; hence, the comments earlier about the lack of effective protocol layering with the T=0 protocol.

An ISO 7816-4 APDU is

- Link-level protocol independent
- Defined at the application level

An instruction APDU is a message structure which carries a command or instruction (and perhaps data) from the reader to the card. A response APDU is a message structure that carries a response (and perhaps data) from the card back to the reader.

The ISO 7816-4 APDU

The messages used to support the ISO 7816-4 defined application protocol(s) comprise two structures—one used by the reader-side of the channel to send commands to the card and the other used by the card to send responses back to the reader. The former is referred to as the *command APDU* and the latter as the *response APDU*.

As illustrated in Figure 4.5, the command APDU comprises a *header* and a *body*, each of which is further subdivided into several fields. The header includes CLA, INS, P1, and P2 fields. CLA and INS define an application class and instruction group as described, for example, in ISO 7816-4. The P1 and P2 fields are used to qualify specific instructions and are therefore given specific definitions by each [CLA,INS] instruction. The body of the APDU is a variable size (and form) component which is used to convey information to the card's APDU processor as part of a command or to convey information from the card back to the reader as part of a response to a command. The Lc field specifies the number of bytes to be transferred to the card as part of an instruction; it contains the length of the data field. The data field comprises data which must be conveyed to the card in order to allow its APDU processor to execute the command specified in the APDU. The Le field specifies the number of bytes which will be returned to the reader by the card's APDU processor in the response APDU for this particular command. The body of the APDU can have four different forms:

- No data is transferred to or from the card, so the APDU includes only the header component. This is referred to as Case 1.

- No data is transferred to the card, but data is returned from the card so the body of the APDU includes only a non-null Le field. This is referred to as Case 2.

- Data is transferred to the card, but no data is returned from the card as a result of the command so the body of the APDU includes the Lc field and the data field. This is referred to as Case 3.

- Data is transferred to the card and data is returned from the card as a result of the command so the body of the APDU includes the Lc field, the data field, and the Le field. This is referred to as Case 4.

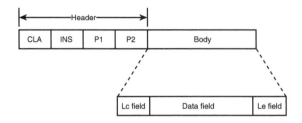

Figure 4.5. *The command APDU structure.*

Figure 4.6 illustrates the much simpler structure of the response APDU structure. It includes a body and a trailer. The body is either null or includes a data field, depending on the specific command that it is a response to, and on whether that command was successfully executed by the card's APDU processor. If the response APDU does include a data field, its length is determined by the Le field of the command to which the response corresponds.

Figure 4.6. *The response APDU structure.*

The response APDU also includes a trailer field, which can comprise two fields of status information that are referenced as SW1 and SW2. These fields return (from the card's APDU processor) to the reader-side application a status code that, according to ISO 7816-4, has a numbering scheme in which one byte is used to convey an error category and the other byte is used to convey a command specific status or error indication. This numbering scheme is illustrated in Figure 4.7. Appendix A includes a table of error/status codes with each command.

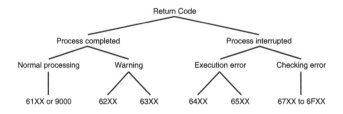

Figure 4.7. *ISO/IEC 7816-4 return codes.*

The CLA code that is included in each command APDU has two additional components to be noted:

- The two low-order bits of the CLA byte can be used to designate a logical communication channel between the reader-side application and the card's APDU processor.

- The next two higher-order bits of the CLA byte can be used to indicate that secure messaging is to be used between the reader-side application and the card's APDU processor.

Once the link-level protocol is established between the reader-side application and the card's APDU processor, a base-level (command) logical channel is created. This is indicated (in the CLA byte) by both of the low-order bits being 0. Additional logical channels can be created by using the `Manage Channel` command, which is defined by ISO 7816-4.

ISO 7816-4 also defines a modest, secure messaging protocol which can be used to ensure privacy and integrity of APDUs transferred between the reader-side application and the card's APDU processor.

The File System API

A central application for smart cards defined by the ISO/IEC 7816-4 standard is a file system. The file system is actually applied to the nonvolatile memory on the smart card; generally EEPROM. The file system defined is a relatively straightforward hierarchical structure comprising three basic elements:

- A master file (MF) component
- A dedicated file (DF) component
- An elementary file (EF) component

The MF component is the root of the file hierarchy; there is only one MF on a smart card. An MF may contain, as elements, a DF, or even many DFs, and it may contain zero to many EFs. The DF component is essentially a container for EF components; a DF may contain zero to many EFs. An EF component may contain only records. This simple hierarchical structure is illustrated in Figure 4.8.

Several characteristics of the smart card file system are significantly different from typical (that is, disk based) file systems. These differences are almost exclusively due to the physical characteristics of the EEPROM memory system, specifically the facts that EEPROM memory can be subjected to only a modest number of erase and write cycles and that it is significantly faster to write to EEPROM memory in a cumulative fashion than in a pure erase and then write fashion. The first of these characteristics resulted in the definition of a rather unique file structure called a *cyclic* file. The second characteristic resulted in rather unique definitions of the various file write commands.

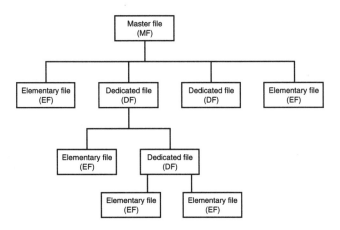

Figure 4.8. *The smart card file system architecture.*

The cyclic file is actually a ring buffer of physical records that are addressed and accessed as a single record. On successive write operations, the next physical record (in the ring of physical records) is accessed. The net result is that erase and write operations can be spread across a wider selection of EEPROM memory locations. This mitigates somewhat the limit (generally on the order of 100,000 cycles) on the number of times that a specific EEPROM memory location can be erased and rewritten.

EEPROM memory has the additional interesting characteristic that it is significantly faster to set additional bits in a byte-sized memory location than it is to erase all the currently set bits and then rewrite them. This fact becomes doubly useful in certain operations (for example, manipulating a purse value on a smart card) where it is required that operations on a file be performed in such a fashion that the values stored in the file are well understood at any point in time, even if power is removed from the smart card in the middle of a write operation. To facilitate the exploitation of these characteristics, the write operations to a smart card file are typically bit set operations while the update operations are actually erase and rewrite operations that we generally associate with file-writing operations. These characteristics will be examined in more detail in the following sections.

Master File Characteristics

Each smart card file system has exactly one master file. The MF serves as the root of the hierarchical file structure. In the parlance of general file systems, the MF is a container or a directory; it may contain other dedicated (or directory) files or it may contain elementary files.

Any file can be identified by a 2-byte file identifier. The file identifier 3F00 is reserved for the MF; that is, there is only one file on the card with a file identifier of 3F00, and the file with that identifier is the MF.

Dedicated File Characteristics

A DF is also a container or a directory file in the same vein as the MF. A DF forms a subdirectory within the file hierarchy rooted in the MF. A DF can also be identified by a file identifier. A DF must be given a unique file identifier within the DF (or MF) that contains it. This allows for the creation of a unique path designation for a file; that is, a path is simply the concatenation of the file identifiers of the file in question and of all the DFs between the file in question and its containing DF or MF.

On some smart cards, a DF can also be referenced by a name which may be from 1 to 16 bytes long. The naming conventions for the DF name are found in the ISO/IEC 7816-5 specification.

Elementary File Characteristics

An EF is a leaf node of the file hierarchy. It is a file that actually contains data. There are two variants of EFs: an internal EF, which is to be used by applications on the card, and a working EF, which is used as a storage mechanism for information used by an off-card application.

Within a specific DF, an EF may be identified by a short (5-bit) identifier. There are four variants of working EFs:

- A transparent file
- A linear, fixed-length record file

- A linear, variable-length record file
- A cyclic, fixed-length record file

A transparent file can be viewed as a string of bytes. When a command is used to read or write information from a transparent file, it is necessary to provide a byte offset (from the start of the file) to the specific byte (within the transparent file) where reading or writing should begin. A command to read or write information from/to a transparent file will also contain a counter or length of the byte string to be read or written to the file.

Fixed- or variable-length record files are, as their name suggests, files that comprise subdivisions called *records*. Records (within a file) are identified by a sequence number. In a fixed-length record file, all the records contain the same number of bytes. In a variable-length record file, each record in the file can contain a different number of bytes. As might be suspected, a variable-length record file generally has a significantly higher overhead in read/write access time and in the amount of administrative (data storage) overhead required by the file system.

A cyclic file is a rather unique (to smart card file systems) structure. It allows applications to access a record in a consistent, transparent fashion and yet have the file system itself map this access into a variety of different physical records. This allows the limits of erase and rewrite cycles on EEPROM memory to be somewhat mitigated.

A cyclic file is best thought of as a ring of records. Each successive write to the file performs the operation on the next physical record in the ring. Read operations are performed on the last physical record to which it was actually written.

To manipulate the smart card file system, an application level protocol is defined in the form of a collection of functions for selecting, reading, and writing files. These functions are discussed qualitatively in the following sections. In Appendix A, these commands are presented in a more quantitative fashion, including the specific CLA and INS designations, along with the error and status codes generally returned by the various commands.

The Select File Command

The Select File command is used to establish what may be thought of as a logical pointer to a specific file in the smart card's file system. Once a file is selected by this command, any subsequent file manipulation commands, such as those to read or write information, will operate on the file pointed to by this logical point-er. Access to the smart card's file system is not multithreaded (from the card's view-point), but it is possible to have multiple such logical pointers in play at any point in time. This is done by using the `Manage Channel` command to establish multi-ple logical channels between the reader-side application and the card. Commands to access different files can then be multiplexed (by the reader-side application), allowing different files on the card to be in various states of access by the reader-side application at the same time.

The primary piece of information the `Select` command must convey (from the reader-side application to the smart cards APDU processor) is the identification of the file that this logical pointer must point to. This identification can be provided in three ways (with the specific addressing mechanism being indicated in the data field of the `Select File` command APDU):

- By file identifier (2-byte value)
- By DF name (string of bytes identifying the DF)
- By path (concatenation of file identifiers)
- By short ID

Not all smart cards support all four naming mechanisms. For example, the 3K Multiflex card included with this book supports only the first mechanism.

The `Read Binary` Command

The `Read Binary` command is used by a reader-side application to retrieve some segment of an EF on the card. The EF being accessed must be a transparent file; that is, it cannot be a record-oriented file. If a `Read Binary` command is attempt-ed on a record-oriented EF, the command will abort with an error indicator being returned by the card to the reader-side application.

Two parameters are passed from the reader-side application to the card for this command: an offset pointer from the start of the file to initial byte to be read, and the number of bytes to be read and returned to the reader-side application.

The `Write Binary` Command

The `Write Binary` command is used by a reader-side application to put informa-tion into a segment of an EF on the card. The file being accessed must be a trans-parent file; that is, it cannot be a record-oriented file. If a `Write Binary` command is attempted on a record-oriented EF, the command will abort with an error indicator being returned by the card to the reader-side application.

Depending on the attributes passed from the reader-side application to the card in the `Write Binary` command, the command can be used to either set a series of bytes in the EF (that is, set selected bits within the designated bytes to a value of 1), clear a series of bytes in the EF (that is, set selected bits within the designated bytes to a value of 0), or do a one-time write of a series of bytes in the EF.

The `Update Binary` Command

The `Update Binary` command is used by a reader-side application to erase directly and store a contiguous sequence of bytes in a segment of an EF on the card. The file being accessed must be a transparent file; that is, it cannot be a record-oriented file. If an `Update Binary` command is attempted on a record-oriented EF, the command will abort with an error indicator being returned by the card to the reader-side application.

The `Update Binary` command provides the functions that one would normally associate with a file write command. That is, a string of bytes provided in the command are actually written into the EF on the card, with those byte positions in the file on the card being erased first. The net result is that the string of bytes found in the designated position within the EF on the card is exactly the string sent by the reader-side application in the `Update Binary` command.

Input parameters for the command include an offset pointer from the start of the file and a byte count of the total number of bytes to be written.

The `Erase Binary` Command

The `Erase Binary` command is used by a reader-side application to clear bytes within an EF on a card. The file being accessed must be a transparent file; that is, it cannot be a record-oriented file. If an `Erase Binary` command is attempted on a record-oriented EF, the command aborts, and an error indicator is returned by the card to the reader-side application.

Two parameters are specified as part of the command: an offset from the start of the EF to the segment of bytes within the file to be erased and the number of bytes within that segment.

The `Read Record` Command

The `Read Record` command is a command sent by a reader-side application to read and return the contents of one or more records in an EF on a card. This command must be executed against a record-oriented EF. If it is applied to a transparent EF, the command will abort and an error indicator will be sent from the card back to the reader-side application.

Depending on the parameters passed through the command, either the one designated record is read and returned, or all the records from the beginning of the file to the designated record are read and returned, or all the records from the designated record to the end of the file are read and returned.

The Write Record Command

The Write Record command is a command sent by a reader-side application to write a record into an EF on the card. This command must be executed against a record-oriented EF. If it is applied to a transparent EF, the command will abort and an error indicator will be sent from the card back to the reader-side application.

As with the Write Binary command, this command can actually be used to achieve one of three results: a one-time write of a record into the EF, setting of specific bits within a specific record in the EF, or clearing of specific bits within a specific record in the EF.

Several addressing shortcuts may be used in this command to specify the record to be written to; including the first record in the EF, the last record in the EF, the next record in the EF, the previous record in the EF, or a specific record (identified by number) within the EF.

The Append Record Command

The Append Record command is a command sent by a reader-side application to either add an additional record at the end of a linear, record-oriented EF on a card or to write the first record in a cyclic, record-oriented EF on a card. If it is applied to a transparent EF, the command will abort and an error indicator will be sent from the card back to the reader-side application.

The Update Record Command

The Update Record command is a command sent by a reader-side application to write a record into an EF on the card. This command must be executed against a record-oriented EF. If it is applied to a transparent EF, the command will abort and an error indicator will be sent from the card back to the reader-side application.

As with the Update Binary command, this command is used to write a specific record into an EF. The net result of the operation is that the specific record in the EF is erased and the new record specified in the command is written into the EF.

The Get Data **Command**

The Get Data command is a command sent by a reader-side application to read and return the contents of a data object stored within the file system on the card. This command tends to be very card-specific. That is, the definition of just what constitutes a data object varies widely from card to card.

The Put Data **Command**

The Put Data command is a command sent by a reader-side application to put information into a data object stored within the file system on the card. This command tends to be card-specific. That is, the definition of what constitutes a data object varies widely from card to card.

The Security API

Associated with each component of the file system is a list of access properties. Through these access properties, a state can be defined such that the smart card system must be put into that state through the successful execution of a series of commands by the reader-side application before that component of the file system can be accessed. At the most basic level, the operations to be performed on the file system are to select a specific file and then write information to that file or read information from that file. As shown later, the access properties may be as simple as requiring the reader to provide a predefined personal identification number (PIN) or as complex as the reader proving that it possesses some shared secret (such as a key) with the card by properly encrypting or decrypting a string of bytes provided by the card. These mechanisms are reviewed in more detail in the following sections.

The Verify **Command**

The Verify command is a command sent by a reader-side application to the security system on the card to allow it to check for a match to password-type information stored on the card. That is, this command is used to allow the reader-side application to convince the card that it (the reader-side application) knows a password maintained by the card to restrict access to information on the card.

The password-type information may be attached to a specific file on the card or to part or all of the file hierarchy on the card. Successful execution of this command indicates that the reader-side application did know the correct password and puts the card into a state such that a subsequent access to a file guarded by this password information will succeed.

If the Verify command fails (that is, the password required by the card is not correctly provided by the reader-side application), an error status indicator is returned by the card to the reader-side application.

The Internal Authenticate Command

The Internal Authenticate command is a command sent by a reader-side application to the security system on the card to allow the card to prove that it possesses a secret key that is shared with the reader-side application. To prepare this command, the reader-side application creates a set of challenge data; that is, essentially the reader-side application generates a random number. This number is then encrypted with some agreed on algorithm (with the card); this constitutes a challenge to the card.

When given the command, the card decrypts the challenge with a secret key stored in a file on the card. The information derived from the decryption is then passed back to the reader-side application as a response to the command. If the card really does have the correct secret key, the information passed back will be the random number generated by the reader-side application prior to issuing the Internal Authenticate command.

This command is used by the reader-side application to authenticate the card's identity. That is, when the command is successfully completes, the reader-side application knows the identity of the card and can give to the card access to information or services within the reader-side application.

The External Authenticate Command

The External Authenticate command is used by a reader-side application in conjunction with the Get Challenge command (described in the next section) to allow the reader-side application to authenticate its identity to the card.

Through the Get Challenge command, the reader-side application receives a set of challenge data from the card (that is, a random number generated by the card). The reader-side application then encrypts this information with a secret key. This then forms a cryptogram that is sent to the card via the External Authenticate command. If the reader-side application knows the same secret key that is stored on the card, then when the card decrypts the cryptogram it will find the same random number generated by the last Get Challenge command. Therefore, the card now knows the identity of the reader-side application and can give it (the reader-side application) access to data stored on the card.

The attractive characteristics of this method (from a security standpoint) is that the secret key used to authenticate identity between the reader-side application and the card was never transferred between the reader-side application and the card.

The Get Challenge Command

The Get Challenge command is used by the reader-side application to extract information that can be used to formulate a cryptogram for the card and validated through an External Authenticate command. The result of this command is the generation of a random number by the card, which is then passed back to the reader-side application.

The Manage Channel Command

The Manage Channel command is used by the reader-side application to open and close logical communication channels between it and the card. When the card initially establishes an application-level protocol with the reader-side application (that is, following the ATR sequence), a basic communication channel is opened. This channel is then used to open or close additional logical channels via the Manage Channel command.

The Envelope Command

The Envelope command supports the use of secure messaging via the T=0 link-level protocol. In secure messaging, the full command APDU should be encrypted. However, since the CLA and INS bytes from the APDU overlay elements of the TPDU, these bytes (in the TPDU) cannot be encrypted if the link-level protocol is still to work correctly. So the Envelope command allows a full APDU to be encrypted and then included in the Envelope command's data section (of its APDU). The card's APDU processor can then extract the "real" command and cause it to be executed.

The Get Response Command

The Get Response command is another command which allows the use of the T=0 link-level protocol for conveying the full range of APDUs. Specifically, the Case 4 type of APDU body cannot be supported with the T=0 protocol. That is, you can't send a body of data to the card and then receive a body of data back as a direct response to that command. For this type of command, by using the T=0 protocol, the initial command results in a response which indicates that more data is waiting (in the card). The Get Response command is then used to retrieve that waiting data.

Note that no other command can be interleaved between the original command and the Get Response command.

Summary

This chapter reviews the two protocol layers through which smart card–aware applications communicate between reader-side components and card-side components. The initial power-up, reset, and ATR sequence establishes a physical (half-duplex) channel between the reader and the card. The card and reader then negotiate a link-level protocol, often resolving to either a T=0 or a T=1 protocol. An application-level protocol is then established on top of the link-layer by using the APDU mechanisms defined in ISO/IEC 7816-4.

Two APIs are examined qualitatively: One provides access to a file system on the smart card and the other provides access to security services on the smart card. Multiple logical channels can be supported between the reader-side application and the smart card's APDU (file or security) processing components. Further, by using cryptographic capabilities, a rudimentary secure messaging facility can be put in place between the reader-side application components and the card-side application components.

A more quantitative review of the functions provided through these APIs is found in Appendix A.

CHAPTER 5

THE SCHLUMBERGER MULTIFLEX SMART CARD

This chapter discusses in depth the 3K Multiflex smart card that comes with this book. In Chapter 10, "The Smart Shopper Smart Card Program," and Chapter 11, "The FlexCash Card: An E-commerce Smart Card Application," we build illustrative smart card applications using this smart card.

The 3K Multiflex smart card contains a 5 MHz Motorola SC21 single-chip 8-bit microcontroller with 6,114 bytes of ROM, 3,008 bytes of EEP-ROM, and 128 bytes of RAM. The chip writes EEPROM 4 bytes at a time. It takes the SC21 7 milliseconds for a 4-byte write operation. The card uses the T=0 communication protocol and can communicate with the outside world at 9600 bps. This is the same card that is used in many smart card programs around the world.

Your Multiflex 3K card conforms to the ISO 7816-4 standard, which is covered in Chapter 4, "Smart Card Commands." Communication between the card and the host uses APDUs, and the EEPROM is organized and accessed as an ISO 7816-4 file system. The card supports 21 commands, including a couple that extend the ISO 7816-4 standard (for example, a create file command that is noticeably missing from the standard).

In this chapter, we will take a detailed tour of your Multiflex 3K card. Starting with how the card responds when it is activated, we describe communication with the card at the bit and byte levels so that you know exactly what is going on. Typically, of course, you will be working on a higher-level API, but it is useful to know what is under the hood, particularly when you have to figure out why something isn't behaving as you think it should.

Activating Smart Cards: Reset and Answer to Reset

The microcontroller on the Multiflex card is reset by lifting the voltage on the card's reset contact for more than 50 microseconds and then lowering it. After conducting some internal consistency checks, the Multiflex operating system on the card selects the master file then transmits its 4-byte answer to reset (ATR) to the terminal on the communication line.

The ATR for the 3K Multiflex card is

$3B_{16}\ 02_{16}\ 14_{16}\ 50_{16}$

and the meaning of each of these bytes is given in Table 5.1.

Table 5.1. The ATR of Schlumberger's 3K Multiflex card.

ISO 7816 Name	Description	Value	Interpretation
TS	Bit synchronization	$3B_{16}$	1 is voltage high.
T0—high-order byte	Other T fields present	0000_2	No other T fields present.
T0—low-order byte	Number of historical bytes present in the ATR	2_{16}	Two historical bytes are present in the ATR.
T1	First historical byte	14_{16}	Component code.
T2	Second historical byte	50_{16}	Mask code.

The first part of the ATR up to the historical byte tail-end is governed by ISO 7816-3. The historical byte section, however, varies widely from card manufacturer to card manufacturer. Schlumberger uses the 2-byte historical tail to identify the microcontroller on the card—the microcontroller itself (component code) and the operating system in the ROM (mask code). A component code of 14_{16} means that the chip is a Motorola SC21. The mask code of 50_{16} means the ROM contains the Multiflex M24E-G2 operating system. Since the SC21 is a 3K part, if you put these two codes together, you get Multiflex 3K.

It would be handy if you could universally identify a smart card by its ATR. Unfortunately, many card manufacturers are secretive about the coding of their ATRs, so it is difficult to write multicard applications whose behavior depends on the ATR without doing so on an ATR-by-ATR basis. After transmitting the ATR, the card goes into a polling loop, running the same instructions over and over, listening for a command from your application program.

Directories and Files

Mostly what you will be doing with a smart card is writing data to it and reading data from it after properly establishing your authorization to do so. First and foremost, a smart card is an active guardian of data. In this section, we cover the organization of data on the Multiflex 3K card. It is an organization you are doubtless already familiar with—directories and files—but there are a couple differences between directories and files on your hard disk and directories and files on a smart card.

Selecting a Directory

Even though the Multiflex operating system makes the master file the current file, since you may want to use other cards with your application, it would be typical for your application program to make this selection again and, in the process, retrieve some information about the specific card with which it was dealing. The master file on all ISO 7816–compliant smart cards has the filename $3F00_{16}$. You can make the master file the current file by sending a 7-byte `Select File` command to the card with the fileId of the master file as its argument. Here's an example:

CO_{16} $A4_{16}$ 00_{16} 00_{16} 02_{16} $3F_{16}$ 00_{16}

The meaning of each of the bytes in the `Select File` command is given in Table 5.2.

Table 5.2. The 3K Multiflex `Select File` command.

ISO 7816 Name	Description	Value	Interpretation
CLA	Class	CO_{16}	
INS	Instruction	$A4_{16}$	
P1	First parameter	00_{16}	
P2	Second parameter	00_{16}	
P3	Third parameter	02_{16}	Two data bytes follow
Data	Data	$3F_{16}$ 00_{16}	FileId of the file to be selected

Whenever you send a command to an ISO 7816 smart card, it will respond with two status bytes called status word 1 (SW1) and status word 2 (SW2). This will tell you what the result of processing your command was. Assuming that there is a file with the file identifier $3F00_{16}$, you'd expect the card to respond with

90_{16} 00_{16}

which is the required ISO 7816-4 status return code for successful completion. This would indicate that the operating system on the card found the file $3F00_{16}$ and that it is now the current file. Instead of $90_{16}\ 00_{16}$ however, the 3K Multiflex card responds with

$$61_{16}\ 14_{16}$$

This is also a successful completion but with some additional information. It says, "I found the file you selected successfully and I have some information about it available for you." The second byte tells how much information is available: 14_{16}, or 20 bytes.

In order to retrieve this file description information, you'll send a `Get Response` command to the card with the number of bytes you want to retrieve as its third argument. In particular, the bytes

$$C0_{16}\ C0_{16}\ 00_{16}\ 00_{16}\ 14_{16}$$

would request the operating system to send back the 20 bytes of information about the master file from your 3K Multiflex card. If you do this for the unused 3K Multiflex card contained in the book, the card returns the following 20 bytes:

$$00_{16}\ 00_{16}\ 0B_{16}\ 10_{16}\ 3F_{16}\ 00_{16}\ 38_{16}\ FF_{16}\ FF_{16}\ 44_{16}\ 44_{16}\ 01_{16}\ 05_{16}\ 03_{16}\ 00_{16}\ 02_{16}$$
$$00_{16}\ 00_{16}\ 00_{16}\ 00_{16}$$

The interpretation is given in Table 5.3.

Table 5.3. File control information for the master file ($3F00_{16}$).

Byte	Description	Value	Interpretation of Value
1–2	Unused	$00_{16}\ 00_{16}$	Unused.
3–4	Free bytes in selected file	$0B_{16}\ 01_{16}$	There are 2,832 bytes available in this directory.
5–6	Fileld of selected file	$3F_{16}\ 00_{16}$	The selected file has fileId $3F00_{16}$.
7	Type of selected file	38_{16}	The selected file is a directory file.
8 High	Byte unused	F_{16}	Unused.
8 Low	Byte unused	F_{16}	Unused.
9 High	Byte access condition for the `Directory` command	F_{16}	The `Directory` command can never be used on this directory.
9 Low	Byte unused.	F_{16}	Unused.
10 High	Byte access condition for the `Delete File` command	4_{16}	You must know a cryptographic key to use the `Delete File` command.
10 Low	Byte access condition for the `Create File` command	4_{16}	You must know a cryptographic key to use the `Create File` command.
11 High	Byte access condition for the `Rehabilitate` command	4_{16}	You must know a cryptographic key to use the `Rehabilitate` command.

Byte	Description	Value	Interpretation of Value
11 Low	Byte access condition for the `Invalidate` command	4_{16}	You must know a cryptographic key to use the `Invalidate` command.
12	Status of the selected file	01_{16}	The file is currently unblocked.
13	Number of bytes in following data	05_{16}	5 bytes of data follow.
14	Features	03_{16}	Unused.
15	Number of subdirectories	00_{16}	There are no subdirectories in this directory.
16	Number of elementary files	02_{16}	There are two elementary files in this directory.
17	Number of secret codes	00_{16}	There are no secret codes in this directory.
18	Unused	00_{16}	Unused.
19	Status of the PIN for this directory	00_{16}	There is no PIN file in the current directory.
20	Status of PIN unblocking key	00_{16}	The PIN unblocking key is not itself blocked.

You can retrieve these 20 bytes of file description information with the Get Response command whenever a directory file is selected. We'll now discuss in detail the meaning of each of the descriptive bytes associated with a directory file.

Bytes three and four say there are 2,832 bytes available for new files and subdirectories in this directory. Since the master file is the root directory on the card and thus all other directories and files must be contained in it, in this case we know there are 2,832 bytes of unused EEPROM space on the whole card. This is the amount of nonvolatile on-card memory available for use by your application. Bytes five and six just repeat the fileId of the selected file.

Byte seven says which of five file types possible on the Multiflex card is selected. The Multiflex smart card supports the five different types of files listed in Table 5.4.

Table 5.4. Multiflex 3K file types.

File Type	Value of File Type Byte	Maximum Record Size	Maximum Number of Records
Directory file	38_{16}		
Transparent file	01_{16}		
Record file with fixed-length records	02_{16}	255 bytes	255
Record file with variable-length records	04_{16}	255 bytes	255
Cyclic file	06_{16}	255 bytes	255

We will discuss the details of these four file types. Files that aren't directory files— binary files, record files with fixed-length records, record files with variable-length records, and cyclic files—are often referred to collectively as *elementary files*. Unstructured binary files are also called *transparent files* because the structure of the file is transparent to the operating system.

Byte eight of the file description is unused for directory files. For elementary files, the high-order 2 bits of the eighth byte restrict the operations that can be applied to the file as shown in Table 5.5.

Table 5.5. Elementary file update access conditions.

Bit 8	Bit 7	Allowed Operations	Disallowed Operations
0	0	Update	Increase, Decrease
0	1	Update, Increase	Decrease
1	0	Update, Decrease	Increase
1	1	Decrease, Increase	Update

The six nibbles (hex digits) of bytes 9 through 11 of the 20 bytes returned by Get Response give the *access conditions* for various operations on the selected file. An access condition states what identity must be established by the entity issuing the command before the command can be executed. For example, an access condition might say that the proper PIN must be presented to the card before a particular file can be read. Associated with each file type is a set of operations that have access conditions associated with them. Table 5.6 shows the file commands that can have access conditions associated with them.

Table 5.6. File operations with access conditions.

Key Protected Operations	Nibble Giving Access Condition
Directory Files	
Directory	9 High
Delete File	10 High
Create File	10 Low
Rehabilitate	11 High
Invalidate	11 Low
Elementary Files	
Read, Seek	9 High
Update, Decrease, Decrease Stamped	9 Low

Key Protected Operations	Nibble Giving Access Condition
Increase, Increase, Stamped	10 High
Create Record	10 Low
Rehabilitate	11 High
Invalidate	11 Low

For example, the value in the high nibble in the tenth byte says what kind of key has to be presented to the card before you can delete a file in the selected directory. There are seven possibilities for values in these single hex digit fields. These values are given in Table 5.7. They describe what authentication operation has to be successfully performed in order to satisfy the access condition and thus be able to perform the command. For example, if 01_{16} is associated with the Read and Seek operation on a particular file, then the cardholder would have to present a valid PIN to the card before the card would allow a Read or Seek on that file.

Table 5.7. Identities or authentications for access conditions.

Key Knowledge Needed	Value of Access Condition Nibble
None—Operation is *always* possible	0_{16}
PIN—4-digit personal identification number	1_{16}
Protected—8-byte cryptographic key	3_{16}
Authenticated—8-byte cryptographic key	4_{16}
PIN and protected	6_{16}
PIN and authenticated	8_{16}
None—Operation is *never* possible	F_{16}

In the file description of the master file on the Multiflex 3K card, the value of the high nibble in the tenth byte is 4_{16}, which means that you have to successfully present an 8-digit cryptographic key to the card before you can delete a file in the master file.

The value of the high nibble in the ninth byte is F_{16} which says that no matter what key you present, you can't use the Directory command; that is, the Directory operation is never possible with the master file. (On the other hand, unlike its 8K brother, the Multiflex 3K card doesn't implement a Directory command, so this access condition is a bit academic.)

Selecting an Elementary File

As indicated in the descriptive data returned from the Get Response command we issued after selecting the master file, there are no subdirectories and two elementary files on an unused Multiflex card. The two elementary files have fileIds 0002_{16} and 0011_{16}. The first elementary file, 0002_{16}, is called the *serial number file* and the second, 0011_{16}, is called the *transport key file*.

The serial number file contains a sequence of 8 bytes that uniquely identifies this card among all the millions of cards ever manufactured by Schlumberger. Like the historical bytes of the ATR, how the serial number is placed on a card varies from manufacturer to manufacturer. The 8 bytes in the serial number file of a Schlumberger card have the following interpretation:

Bytes 1–4	Series number
Byte 5	Customer Identification Code
Bytes 6–7	Schlumberger Manufacturing Site
Byte 8	Usage

The 8 bytes in file 0002_{16} in the 3K Multiflex sitting in the author's computer right now are

$$00_{16}\ 00_{16}\ 0E_{16}\ 67_{16}\ 01_{16}\ 00_{16}\ 00_{16}\ 02_{16}$$

This is card #3687 made for the customer with identification code 1 (Schlumberger itself) at Schlumberger's Pont Audemer factory, and it is a sample card. Schlumberger guarantees that the 8 bytes taken together uniquely identify the card.

The transport key is a key that locks the card while it is being shipped from Schlumberger to you. This way, if somebody breaks into the truck and steals the cards, they aren't in possession of a whole bunch of valid cards from your smart card program. Schlumberger sends you the transport key for your cards via a channel different than the truck. When the cards arrive, you will use the transport key to unlock the cards, to personalize them, and to add new keys to them. At the end of this process, you will overwrite or completely erase the transport key. By the way, the transport key on your 3K Multiflex card is

$$47_{16}\ 46_{16}\ 58_{16}\ 49_{16}\ 32_{16}\ 56_{16}\ 78_{16}\ 40_{16}$$

but don't tell anybody.

A simple transport key is sufficient for relatively low-value cards. Higher-value cards use more elaborate transport key protocols. For example, there may be a different transport key on each card which is a secret function of the serial number of the card (a *diversified key*), or the card may have to receive a properly encrypted version of a challenge it issues to a *mother card* or *batch card* before it unlocks itself.

If we again select the serial number file, 0002, using the Select File command:

$CO_{16} A4_{16} 00_{16} 00_{16} 02_{16} 00_{16} 02_{16}$

we'll get a returned status code of

$61_{16} 0F_{16}$

which means there are 15 ($0F_{16}$) bytes of descriptive information about the serial number file waiting on the card for us. So, we send the card a Get Response to get this information:

$CO_{16} CO_{16} 00_{16} 00_{16} 0F_{16}$

and it returns this:

$00_{16} 00_{16} 00_{16} 08_{16} 00_{16} 02_{16} 01_{16} 00_{16} 04_{16} FF_{16} FF_{16} 01_{16} 01_{16} 00_{16} 00_{16}$

The meaning of the bytes returned from issuing a Get Response after selecting an elementary file is similar to but not exactly the same as the meaning of the bytes returned after selecting a directory file. The meaning of the bytes is the same, no matter what type of elementary file is selected: a transparent file with fixed-length records or a file with variable-length records. See Table 5.8.

Table 5.8. File control information for the serial number file (0002_{16}).

Byte	Description	Value	Interpretation of Value
1–2	Unused	$00_{16} 00_{16}$	Unused.
3–4	Free bytes in selected file	$00_{16} 08_{16}$	There are 8 bytes in this file.
5–6	File ID of selected file	$00_{16} 02_{16}$	The selected file has file ID 0002_{16}.
7	Type of selected file	01_{16}	The selected file is a transparent file.
8 High	Restriction of Update, Increase, and Decrease commands	0_{16}	Only the Update command can be used.
8 Low	Unused	0_{16}	Unused.
9 High	Access condition for Read and Seek commands	0_{16}	Anyone can use the Read and Seek commands on this file.
9 Low	Access condition for Update, Decrease, and Decrease Stamped commands	4_{16}	You must know a cryptographic key to update this file. You can't use Decrease or Decrease Stamped due to byte 8.
10 High	Access condition for Increase and Increase Stamped commands	F_{16}	These commands can never be used on this file.
10 Low	Access condition for the Create Record command	F_{16}	These commands can never be used on this file.

91

continues

Table 5.8. continued

Byte	Description	Value	Interpretation of Value
11 High	Access condition for `Rehabilitate` command	F_{16}	These commands can never be used on this file.
11 Low	Access condition for `Invalidate` command	F_{16}	These commands can never be used on this file.
12	Status of the selected file	01_{16}	The file is currently unblocked.
13	Number of bytes in following data	01_{16}	One byte of data follows.
14	Unused	00_{16}	Unused.
15	Length of record in fixed-length record files	00_{16}	Not a record structure file.

Since we don't have to know any key to read the contents of this file, use a `Read Binary` command to take a look at the 8 bytes in this file. The `Read Binary` command reads bytes from transparent files. The bytes we'll send to the card are

$$CO_{16}\ BO_{16}\ 00_{16}\ 00_{16}\ 08_{16}$$

CO_{16} is the class code and BO_{16} is the instruction code for the `Read Binary` command. The next two bytes say at what offset from the first byte in the file the read should start and the last byte says how may bytes should be read. Since we want to see all the bytes in the file we'll read 8 bytes starting at an offset of 0.

The 8 bytes we get back are

$$00_{16}\ 00_{16}\ 0E_{16}\ 67_{16}\ 01_{16}\ 00_{16}\ 00_{16}\ 02_{16}$$

The first 4 bytes are the serial number of the card ($00_{16}\ 00_{16}\ 0E_{16}\ 67_{16}$) and the second 4 bytes are a manufacturer's code. Taken together, these 8 bytes are guaranteed to be a unique serial number for the smart card.

If we select the transport key file, 0011:

$$CO_{16}\ A4_{16}\ 00_{16}\ 00_{16}\ 02_{16}\ 00_{16}\ 11_{16}$$

and get back the status code:

$$61_{16}\ 0F_{16}$$

and use the `Get Response` command to get the 15 bytes of information about it:

$$00_{16}\ CO_{16}\ 00_{16}\ 00_{16}\ 0F_{16}$$

we get this:

$$00_{16}\ 00_{16}\ 00_{16}\ 26_{16}\ 00_{16}\ 11_{16}\ 01_{16}\ 00_{16}\ F4_{16}\ 40_{16}\ F4_{16}\ 01_{16}\ 01_{16}\ 00_{16}\ 00_{16}$$

The interpretation of these bytes is given in Table 5.9.

Table 5.9. File control information for the external authentication key file (0011_{16}).

Byte	Description	Value	Interpretation of Value
1–2	Unused	$00_{16}\ 00_{16}$	Unused.
3–4	Free bytes in selected file	$00_{16}\ 26_{16}$	There are 38 bytes in this file.
5–6	FileId of selected file	$00_{16}\ 11_{16}$	The selected file has fileId 0011_{16}.
7	Type of selected file	01_{16}	The selected file is a transparent file.
8 High	Restriction of Update, Increase, and Decrease commands	0_{16}	Only the Update command can be used.
8 Low	Unused	0_{16}	Unused.
9 High	Access condition for Read and Seek commands	F_{16}	Nobody can use the Read and Seek commands on this file.
9 Low	Access condition for Update, Decrease, and Decrease Stamped commands	4_{16}	You must know a cryptographic key to update this file. You can't use Decrease or Decrease Stamped because of byte 8.
10 High	Access condition for Increase and Increase Stamped commands	4_{16}	You'd have to know a cryptographic key to use these commands but you can't due to byte 8.
10 Low	Access condition for the Create Record command	0_{16}	Anybody can use this command on this file.
11 High	Access condition for Rehabilitate command	F_{16}	Rehabilitate can never be used on this file.
11 Low	Access condition for Invalidate command	4_{16}	You have to know a cryptographic key to invalidate this file.
12	Status of the selected file	01_{16}	The file is currently unblocked.
13	Number of bytes in following data	01_{16}	One byte of data follows.
14	Unused	00_{16}	Unused.
15	Length of record in fixed-length record files	00_{16}	

If you ignore the access conditions on this file and try to read the contents without successfully presenting the Authenticate key to the card, the status return from the card will be

$$69_{16}\ 82_{16}$$

which means the access condition is not fulfilled for the requested operation. You need to know the key in the transport key file in order to perform useful operations on the card (such as creating new files), but since you can't read the file, it seems like you are stuck.

As described above, the reason the key in this file is called the *transport key* is that it is the key that locks the card during transport from the card manufacturer to you. This prevents somebody from breaking into the box containing the cards and possibly, unknown to you, putting something nasty on the cards. The transport key is typically given to you "out of band," that is by post, fax, telephone, bonded courier, or some way other than how the cards are shipped.

When you receive the cards from the card manufacturer, you will present the transport key to each card and then build the files that describe your application on the card. This process of building a particular application on a generic card is called card *personalization*. The personalization process might also include writing specific data into the files you create on the card, such as the account number with which the card is associated. The personalization process will most likely overwrite the transport key or delete the transport key file altogether.

Since the card included with this book is for learning and experimentation, we aren't going to require you to get in touch with the authors to get the transport key for your card. We're just going to tell it to you here. Just in case you missed it previously, here it is again:

$$47_{16}\ 46_{16}\ 58_{16}\ 49_{16}\ 32_{16}\ 56_{16}\ 78_{16}\ 40_{16}$$

Keys and Key Files

There are two kinds of keys used in the Multiflex smart card: PIN codes and cryptographic keys. PIN codes are used for card-to-person authentication. Cryptographic keys are used for card-to-computer authentication.

PIN codes are used by the card to make sure that the person trying to use the card is authorized to do so. PIN codes are usually four digits long, but they can be up to eight digits long. In a typical scenario, the cardholder is asked to enter a PIN code on a keypad or keyboard attached eventually to the card reader containing the card. The entered value is then sent to the card using the Verify PIN command. If the entered PIN agrees with the value found in the current PIN file, then the access level on the card is set to CHV (which stands for *cardholder verified*) and the cardholder can go ahead and perform all operations authorized to the CHV access condition.

Cryptographic keys are used by the card to authenticate and be authenticated by the terminal or computer into which the card has been inserted. Authentication is performed by demonstrating knowledge of a cryptographic key. There are four ways cryptographic keys are used:

- Verify key—The terminal can demonstrate knowledge of a cryptographic key by simply sending the cryptographic key to the card.

- External authentication—The terminal can demonstrate knowledge of a cryptographic key by using the key to encrypt a challenge provided to it by the card.

- Internal authentication—The card can demonstrate knowledge of a cryptographic key by using the key to encrypt a challenge provided to it by the terminal.

- Protected-mode commands—The terminal can use a command protected by a cryptographic key by sending a challenge encrypted with the key along with the command. (See the section "Protected-Mode Commands" later in this chapter for a full explanation of this use.)

External authentication is a better way for a terminal to authenticate itself to the card than to verify the key because the key itself doesn't pass over the communication line between the terminal and the card.

Each directory on a 3K Multiflex card can contain up to three key files that control access to the files in that directory. A particular directory need not contain any of these key files. It can contain only one, just two, or all three of them. The key files are all transparent elementary files that have special reserved names. The names of the special key files and what each file contains are shown in Table 5.10.

Table 5.10. Standard key files on the 3K Multiflex card.

Key File Identifier	Key File Name(s)	Key File Contents	Maximum Number of Keys
0000_{16}	PIN file	PIN code.	1
0001_{16}	Internal authentication file	Internal cryptographic keys: Keys used by the card to prove its identity to entities outside itself.	16
0011_{16}	External authentication file	External cryptographic keys: Keys the card uses to authenticate entities outside itself.	16

If a PIN code file 0000_{16} is present in the directory, it contains a sequence of 23 bytes describing one PIN code. The format of this descriptor is given in Table 5.11.

Table 5.11. Format of a PIN file.

Byte Number	Description	Sample Values	Interpretation of Sample Values	Comment
1	Activation Byte	FF_{16}	PIN is unblocked	A value of 00_{16} means the PIN is blocked.
2				Reserved for future use.
3				Reserved for future use.
4–11	PIN Code	$01_{16}\ 02_{16}$ $03_{16}\ 04_{16}$ $FF_{16}\ FF_{16}$ $FF_{16}\ FF_{16}$	PIN is 1234	A value of FF_{16} means the byte is ignored in checking a presented PIN.
12	Attempts Allowed	03_{16}	Three sequential incorrect presentations of the PIN will block the PIN.	When the PIN is blocked you must use the Unblocking PIN and the Unblock PIN command to unblock it.
13	Attempts Remaining	03_{16}	Three more attempts to enter the PIN may be made before it is blocked.	A successful presentation resets this value to Attempts Allowed.
14–21	Unblocking PIN Code	$08_{16}\ 07_{16}$ $06_{16}\ 05_{16}$ $04_{16}\ 03_{16}$ $02_{16}\ 01_{16}$	87654321 is the PIN code needed to unblock this PIN.	This key is usually known to the card issuer but not the cardholder. The cardholder must present the card to the card issuer to get it unblocked.
22	Unblock attempts allowed	03_{16}	Three sequential incorrect presentations of the unblocking key will block the PIN forever.	There is no way to unblock a blocked unblocking key, so once it is blocked the PIN is blocked forever.
23	Unblock attempts remaining	03_{16}	Three more attempts at entering the unblocking key may be made before it is blocked forever.	

The two cryptographic key files, 0001_{16} and 0011_{16}, both have the structure given in Table 5.12.

Table 5.12. Format of internal and external authentication key files.

Byte Number	Description	Sample Values	Interpretation of Sample Values	Comment
1	Unused			
2	Length of key 0	8_{16}	Key 0 is 8 bytes long.	Cryptographic keys can be from 1 to 255 bytes long.

Byte Number	Description	Sample Values	Interpretation of Sample Values	Comment
3	Algorithm for key 0	0_{16}	Use DES with key 0.	
4–11	Key 0	FF_{16} FF_{16} FF_{16} FF_{16} FF_{16} FF_{16} FF_{16} FF_{16}		
12	Maximum attempts for key 0	03_{16}	Block key after three successive failed attempts.	
13	Remaining attempts for key 0	03_{16}	There are three failures left before the key is blocked.	
14	Length of key 1	8_{16}	Key 1 is 8 bytes long.	
15	Algorithm for key 1	0_{16}	Use DES with key 1.	
16–23	Key 1	47_{16} 46_{16} 58_{16} 49_{16} 32_{16} 56_{16} 78_{16} 40_{16}		
24	Maximum attempts for key 1	03_{16}	Block key after three sequential failed attempts.	
25	Remaining attempts for key 1	03_{16}	There are three failures left before the key is blocked.	
26	Length of key 2	8_{16}	Key 2 is 8 bytes long.	
27	Algorithm for key 2	0_{16}	Use DES with key 2.	
28–35	Key 2	FF_{16} FF_{16} FF_{16} FF_{16} FF_{16} FF_{16} FF_{16} FF_{16}		
36	Maximum attempts for key 2	03_{16}	Block key after three sequential failed attempts.	
37	Remaining attempts for key 2	03_{16}	There are three failures left before the key is blocked.	
38	Flag for last key	0_{16}	There are more keys in this file.	

Even though you can't read what's in 0011_{16}, we will tell you that the content of the transport key file of your Multiflex card is exactly what is in the Sample Values column in Table 5.12. In other words, cryptographic key 1 for the root directory of your 3K Multiflex card is

$$47_{16} \ 46_{16} \ 58_{16} \ 49_{16} \ 32_{16} \ 56_{16} \ 78_{16} \ 40_{16}$$

Creating a PIN File and Updating the External Authentication Key File

Adding a PIN file to the Multiflex card will require us to use the `Create File` command in the root directory, and this in turn requires Authenticated privileges, so we will have to start out by authenticating ourselves. Since we don't think there are any malicious hackers lurking on the serial connection between our laptop and the smart card reader, we will use the `Verify Key` command rather than the `External Authentication` command to achieve authenticated status on the card.

With the `Verify Key` command, we give the key number in the external authentication file that we want to use to authenticate ourselves to the card along with the key itself. Here's the `Verify Key` command that is sent to the card to achieve Authenticated status:

CLS	INS	P1	Key Number	Key Length	Key
$F0_{16}$	$2A_{16}$	00_{16}	01_{16}	08_{16}	$47_{16} \ 46_{16}$ $58_{16} \ 49_{16}$ $32_{16} \ 56_{16}$ $78_{16} \ 40_{16}$

The card responds with

$$90_{16} \ 00_{16}$$

so we know we have successfully logged in.

Now we issue the `Create File` command to actually create the PIN file:

CLS	INS	Initialize	No. of Recs	Data Length	Unused	Size
$F0_{16}$	$E0_{16}$	00_{16}	FF_{16}	10_{16}	$FFFF_{16}$	0017_{16}

FID	File Type	Access Levels	Status	Length	Access Keys
0000_{16}	01_{16}	$3_{16}F_{16}4_{16}$ $4_{16}F_{16}F_{16}$ $4_{16}4_{16}$	01_{16}	03_{16}	$1_{16}1_{16}F_{16}$ $F_{16}1_{16}1_{16}$

Next, because we want to write the key values (PIN code and unblocking key) into the file, we select the file we just created:

CO_{16} $A4_{16}$ 00_{16} 00_{16} 02_{16} 00_{16} 00_{16}

We issue a `Get Response` command to retrieve the description of the file just to make sure that everything is okay:

CO_{16} CO_{16} 00_{16} 00_{16} $0F_{16}$

And we get back

00_{16} 00_{16} 00_{16} 17_{16} 00_{16} 00_{16} 01_{16} $3F_{16}$ 44_{16} FF_{16} 44_{16} 01_{16} 01_{16} 00_{16} 00_{16}

which you can read like your name by now.

Everything looks fine, so we'll go ahead and write the PIN code, the unblocking key, and their attempt parameters into the file using the `Update Binary` command:

CLS	INS	Offset High	Offset Low	No. Bytes to Write	Data to Write
CO_{16}	$D0_{16}$	00_{16}	00_{16}	17_{16}	01_{16} FF_{16} FF_{16} 31_{16}
					32_{16} 33_{16} 34_{16} FF_{16}
					FF_{16} FF_{16} FF_{16} $0F_{16}$
					$0F_{16}$ 31_{16} 32_{16} 33_{16}
					34_{16} 35_{16} 36_{16} 37_{16}
					38_{16} $0F_{16}$ $0F_{16}$

The 23 bytes written into the PIN file exactly match the description of the PIN file given in Table 5.11.

If we now read back the 23 bytes we put into the PIN file

CO_{16} $B0_{16}$ 00_{16} 00_{16} 17_{16}

we get this:

01_{16} FF_{16} FF_{16} 31_{16} 32_{16} 33_{16} 34_{16} FF_{16} FF_{16} FF_{16} FF_{16} $0F_{16}$ $0F_{16}$ 31_{16} 32_{16} 33_{16} 34_{16} 35_{16} 36_{16} 37_{16} 38_{16} $0F_{16}$ $0F_{16}$

This is exactly what we wanted. The PIN code is 1234, the unblocking key is 12345678, and both will take 15 sequential failures before blocking. Notice that by making the trailing 4 bytes of the PIN code FF_{16}, we have indicated that this is only a 4-digit PIN, since the default fill byte for keys is FF_{16}. The `Verify` PIN command:

CLA	*INS*	*P1*	*Key Number*	*Key Length*	*PIN*
$C0_{16}$	20_{16}	00_{16}	01_{16}	08_{16}	$31_{16}\,32_{16}\,33_{16}\,34_{16}\,FF_{16}$ $FF_{16}\,FF_{16}\,FF_{16}$

now returns this:

$90_{16}\,00_{16}$

While we're at it, let's go ahead and use the `Update Binary` command to add keys 0 and 2 into the external authorization key file, 0011_{16}, in the root directory. Set key 0 to `SCDK1997`:

$C0_{16}\,D6_{16}\,00_{16}\,01_{16}\,0C_{16}\,08_{16}\,00_{16}\,53_{16}\,43_{16}\,44_{16}\,4B_{16}\,31_{16}\,39_{16}\,39_{16}\,37_{16}$ $05_{16}\,05_{16}$

and key 2 to `37743128`:

$C0_{16}\,D6_{16}\,00_{16}\,19_{16}\,0C_{16}\,08_{16}\,00_{16}\,33_{16}\,37_{16}\,37_{16}\,34_{16}\,33_{16}\,31_{16}\,32_{16}\,38_{16}$ $07_{16}\,07_{16}$

The first key allows 5 failed attempts before blocking and the second key allows 7. Obviously, you can use these commands with your own key values to set and reset the keys on your own card.

Record Files and Seek

The key files are transparent files. They have no internal structure and you read and write them as a long sequence of bytes. A record file or cyclic file is organized as a series of records. All the records in a record file with fixed-length records and in a cyclic file are of the same length. Obviously, the records in a record file with variable-length records can be of different lengths. There can be up to 255 records in a record or cyclic file and each record can be up to 255 bytes long.

You are no doubt familiar with record-oriented files on regular computer file systems. Records are also known as TAB cards or rows in tables or, in the most modern parlance, objects. In a typical application, each record contains all the information about a particular item such as a person in an organization, a product in an inventory, or a book in a library.

One of the functions that you frequently perform on a record file is to search for a particular record. The utility of record files, at least large record files, on a smart card would be greatly reduced if one were to have to read all the records from a record file in a smart card from the terminal to find a particular record. It would take a considerable amount of time to do this. Fortunately, the Multiflex card has a Seek instruction that will carry out this search on the card by using the card's processor.

Consider a record file that contains the following 10 records:

```
Sally Green
Ted Yellow
Bobby Blue
George Gray
Barbara Bloom
Annette Anise
Steve Steamboat
Gary Grime
Suzie Creamcheese
Lisa Lavender
```

and assume we have selected this file so that it is the current file.

The Multiflex Seek instruction lets us search the records in this file with a particular pattern, starting at a given offset in each record and either starting from the beginning of the file or from the current position in the file. If the pattern is found in a record, then the card returns success and the found record becomes the current record. Otherwise, if a record containing the pattern at the specified offset is not found, an error condition is returned.

As an example, to find Gary Grime's record, we would send the following command to the card:

CLA	INS	Offset	Mode	Data Length	Pattern
$F0_{16}$	$A2_{16}$	00_{16}	00_{16}	$0A_{16}$	$47_{16}\ 61_{16}\ 72_{16}\ 79_{16}\ 20_{16}$ $47_{16}\ 72_{16}\ 69_{16}\ 6D_{16}\ 65_{16}$

If we wanted to find all the Garys, we would issue the command

CLA	INS	Offset	Mode	Data Length	Pattern
$F0_{16}$	$A2_{16}$	00_{16}	00_{16}	04_{16}	$47_{16}\ 61_{16}\ 72_{16}\ 79_{16}$

which would find the first Gary and then repeat the command

CLA	INS	Offset	Mode	Data Length	Pattern
$F0_{16}$	$A2_{16}$	00_{16}	02_{16}	$0A_{16}$	$47_{16}\ 61_{16}\ 72_{16}\ 79_{16}$

until we got the Not Found error. This would find all the other Garys. At any time, we could use the card's Read Record command to retrieve the Gary record that was found.

Cyclic Files and Electronic Purses

The difference between a record file with fixed-length records and a cyclic file is that the cyclic file "wraps around" and the record file doesn't. For example, when you try to read the next record after the last record in a record file, you get an error, whereas with a cyclic file you get the first record in the file. Similarly, if you try to read the record previous to the first record in a record file, you get an error, whereas with a cyclic file you get the last record in the file.

This wrap-around quality of cyclic files makes them handy for transaction logs. Imagine, for example, that the current record in a cyclic file contains the amount of money currently in an electronic purse, and the cardholder uses the smart card to buy something. To keep track of the transactions on the purse, just subtract the purchase price from the amount in the current record and write the result into the previous record. This will automatically make the previous record the current record in the cyclic file and thus ready it for the next transaction. If there are N records in the cyclic file, then the last N transactions are remembered in the electronic purse's transaction file. If something goes wrong with the writing of a transaction, the purse can easily be backed up to the transaction before the garbled one and the owner of the value in the purse is only out one transaction.

Two commands in the Multiflex card automatically perform this electronic purse functionality using a cyclic file: `Decrease` and `Increase`. As must be clear, `Decrease` decreases the amount in the purse and `Increase` increases the amount in the purse.

Suppose we have selected a cyclic file that we are using as an electronic purse and have been authenticated sufficiently to use the `Increase` command on this file. If we send the following `Increase` command to the card

CLA	INS	P1	P2	Data Length	Pattern
$F0_{16}$	32_{16}	00_{16}	00_{16}	03_{16}	000010_{16}

then 000010_{16} will be added to the value found in the rightmost 3 bytes in the current record. The resulting 3 bytes will be written into the rightmost 3 bytes of the record previous to this record, and finally this previous record will be made the current record. The `Decrease` command works the same way except that the value is subtracted from rather than added to the value found in the current record.

The `Increase` command will not attempt to increment the value in the electronic purse above $FFFFFF_{16}$ and the `Decrease` command will not attempt to decrement the value below 000000_{16}. If the command with the value it contains would cause either of these situations, it is not executed and an error condition is returned. The arithmetic of the 3K Multiflex electronic purse is obviously integer arithmetic, so the units of the purse have to be defined appropriately for the application at hand—dollars, cents, pounds, pence, francs, centimes, yen, lira, and so on.

The utility of those upper 2 bits in the high nibble of byte 8 of the data associated with the `Create File` command is now apparent. By appropriately setting these bits, the program designer can ensure that the electronic purse can only be decremented or can only be incremented.

Now suppose the terminal is a vending machine that is going to decrement the purse and, in return for removed value, release a product to the cardholder. If all the vending machine gets to see is the $90_{16} \, 00_{16}$ return from its `Decrease` command, it has no way of knowing if there really is an authentic purse out there that actually has been decremented or if there is just a laptop computer that is sending back the satisfactory completion status. A $90_{16} \, 00_{16}$ status code is weak evidence on which to release a product.

There are a number of techniques used to counter this threat. The big brother of your 3K Multiflex smart card, the 8K Multiflex card, has *stamped* variants of the `Increase` and `Decrease` commands to counter this threat by letting the terminal know it is dealing with an authentic smart card. Stamped variants of the commands work just like the unstamped versions, but in addition they return information via `Get Response` that lets the terminal check if the operation was performed by an authentic card. The information returned by a `Stamped Increase` or `Stamped Decrease` command is

- The new value of the purse
- The amount added or subtracted from the old value
- A cryptogram

The new value of the purse information is of use primarily in online situations. It can be combined with a reading of the serial number of the card to keep a central record of cards and current values that could be checked and updated over the network.

The cryptogram is of use primarily in offline situations, such as our hypothetical vending machine. Before the `Stamped Increase` or `Stamped Decrease` command is sent to the card, the vending machine sends a sequence of 8 random bytes to the card. The card is expected to return an encryption of these 8 bytes upon successful completion of the increment or decrement. If the cryptogram returned by the card does not decrypt using a key possessed by all authentic cards, then the vending machine knows that a valid electronic purse has not been decremented and thus will not release a product.

The reason the random string of bytes is used in the protocol is to prevent replay attacks. If a fixed string were used, then an attacker could record the byte stream returned by an authentic card and simply feed that string back to the terminal with the amount adjusted to convince the terminal that an authenticated card had been decremented.

Multiflex Commands

Earlier sections describe some of the commands to which the Multiflex card responds (for example, Select File, Get Response, and Read Binary). The Multiflex smart card recognizes and responds to a total of 21 commands. These are the basic building blocks you would use to construct a Multiflex smart card application. Table 5.13 lists all 21 3K Multiflex commands.

Table 5.13. The 21 3K Multiflex commands.

Command Name	Command Description	Access Conditions	Protected Mode
Change PIN	Change the PIN in the selected PIN file.	None	
Create File	Create a new file or subdirectory in the current directory.		Yes
Create Record	Create a new record in the selected record file.		Yes
Decrease	Make the next record in the selected cyclic file the current record minus the given value.		Yes
Delete File	Delete a file from the selected directory.		Yes
External Authentication	Retrieve encrypted challenge from terminal and check.	None	
Get Challenge	Retrieve a challenge from the card	None	
Get Response	Retrieve information about a selected directory or file.	None	
Increase	Make the next record in the selected cyclic file the current record plus the given value.		Yes
Internal Authentication	Retrieve the encryption of the challenge from the card.		Yes
Invalidate	Completely block access to the selected file.		Yes

Command Name	Command Description	Access Conditions	Protected Mode
Read Binary	Read data from a section of the selected transparent file.		Yes
Read Record	Read a record from the selected linear elementary file.		Yes
Rehabilitate	Remove the block on the selected file.		Yes
Seek	Search the selected record file for records containing a given byte string.		
Select	Make the file with the given file identifier the selected file.	None	
Unblock PIN	Unblock a blocked PIN.	None	
Update Binary	Overwrite data in a section of the selected transparent file.		Yes
Update Record	Overwrite a record in the selected linear elementary file.		Yes
Verify PIN	Present a PIN to the smart card.	None	
Verify Key	Present a cryptographic key to the smart card.	None	

All the commands in Table 5.13 are described in detail in the 3K Multiflex documentation on the CD-ROM.

Protected-Mode Commands

Stamped Increase and Decrease commands are examples of a technique a terminal can use to guard against rogue cards. Protected-mode commands are how a card protects itself against rogue terminals.

In order for the terminal to execute a command that has been given a protected-mode access condition, the terminal must first get a random eight challenge from the card using the Get Challenge command. The terminal encrypts this challenge using a specific key shared with the card and returns the encrypted challenge—the cryptogram—along with the protected mode command. The card decrypts the cryptogram and if it gets the challenge it sent, then it knows the terminal possesses the same key as it does and it executes the command. If the cryptogram does not successfully decrypt, the card aborts the command and returns an error status.

Suppose the Update command has been given a protected mode access condition on a selected transparent file using key 1 in the external authentication key file and the terminal wants to write ab into the first 2 bytes of the file. The terminal begins by sending the card a Get Challenge command. This is to get the plain text of the cryptogram it has to return to the card with the Update command:

CLA	INS			Data Length
$C0_{16}$	$D6_{16}$	00_{16}	00_{16}	08_{16}

The card responds with

$64_{16}\ 46_{16}\ 27_{16}\ E0_{16}\ 07_{16}\ 9D_{16}\ D8_{16}\ 6C_{16}\ 90_{16}\ 00_{16}$

which is 8 random bytes followed by the normal completion status return. The terminal DES encrypts these with key 1 in the external authentication key file (47_{16} $46_{16}\ 58_{16}\ 49_{16}\ 32_{16}\ 78_{16}\ 40_{16}$) and sends the card the Update command together with this encryption:

CLA	INS	Offset High	Offset Low	Length	Cryptogram
$C0_{16}$	84_{16}	00_{16}	00_{16}	$0A_{16}$	$61_{16}\ 62_{16}\ 0D_{16}\ 31_{16}$ $A8_{16}\ F3_{16}\ 1C_{16}\ EF_{16}$ $78_{16}\ F8_{16}$

Here $61_{16}\ 62_{16}$ is the character sequence ab that the command will write starting at offset 0000_{16} in the file and $0D_{16}\ 31_{16}\ A8_{16}\ F3_{16}\ 1C_{16}\ EF_{16}\ 78_{16}\ F8_{16}$ is the DES encryption of the challenge.

The card decrypts the trailing 8 bytes using the key associated with the Update command in the current external authentication key file, and upon getting what it sent in response to the Get Challenge command, which it remembered, executes the Update command.

Internal and External Authentication

As you may have already gathered, the way you establish your identity in the world of smart cards is by demonstrating possession of a secret—typically a key of some sort. You can demonstrate such a possession by simply showing the secret, as in the case of a PIN, or you can demonstrate possession of it by doing something that only somebody with the secret could do, such as encrypting a message with the key.

Internal and external authentication use the latter technique to establish identity. The terminal uses internal authentication to establish the identity of the card and the card uses external authentication to establish the identity of the terminal.

Both protocols start by wondering about the other side's identity and sending a challenge—a random 8-byte sequence—to the side whose identity is being questioned. The side receiving the challenge encrypts it with its secret, a particular key, and sends back the encrypted result. The side that originally sent the challenge decrypts the encrypted message using the key it knows the other possesses if it is authentic. If the message decrypts satisfactorily (that is, the decrypted challenge is identical to the challenge originally sent), then the side sending the original challenge knows that the other side possesses a particular key and this establishes its identity. What is critical is to note this possession has been demonstrated without exposing the secret itself.

Since we've already seen two examples of the card establishing the identity of the external world, `Verify PIN` and `Verify Key`, let's walk through an example of the terminal establishing the identity of the card. A terminal would do this to ensure that it isn't unwittingly talking to a rogue smart card; that is, a smart card that looks like the real thing to the terminal but isn't.

We started by creating an internal authentication key file that contains the 8-byte key 0 Vagabond. This file, 0001_{16}, is in the root directory and looks as shown in Table 5.14.

Table 5.14. Internal authentication file.

Byte Number	Description	Value	Interpretation of Value
1	Unused	00_{16}	Unused
2	Length of key 0	8_{16}	Key 0 is 8 bytes long.
3	Algorithm for key 0	0_{16}	Use DES with key 0.
4–11	Key 0	$56_{16}\ 61_{16}\ 67_{16}$ $61_{16}\ 62_{16}\ 6F_{16}$ $6E_{16}\ 64_{16}$	Vagabond
12	Maximum attempts for key 0	03_{16}	Block key after three successive failed attempts.
13	Remaining attempts for key 0	03_{16}	There are three failures left before the key is blocked.
14	Length of key 1	0_{16}	No more keys in the file.

The terminal starts the authentication process by sending to the card an `Internal Authentication` command, which contains the random challenge

CLA	INS	P1	P2	Length	Challenge
$C0_{16}$	88_{16}	00_{16}	00_{16}	08_{16}	$2A_{16}\ 61_{16}\ BD_{16}\ 80_{16}\ A4_{16}\ F9_{16}\ F3_{16}$ $3E_{16}$

The card encrypts the challenge using the key and algorithm of key 0 in 0001_{16} and then responds with the status code

$61_{16}\ 06_{16}$

which means "OK, I encrypted the challenge with key 0 in the file 0001 and I have six bytes of the result to give back to you."

The terminal uses the `Get Response` command to get the 6 bytes from the card

$$1E_{16}\ 86_{16}\ 15_{16}\ C7_{16}\ 15_{16}\ 8C_{16}$$

and compares them with the first 6 bytes of its own DES encryption of the challenge using the key `Vagabond`. The bytes are identical, so the card possesses the key that the terminal was looking for, hence the identity of the card has been established.

Authentication States and Authentication State Transitions

We have said that there can be a PIN and up to 16 external authentication keys in each directory on the card. We have also noticed that the access conditions for operations on files are expressed in terms of a PIN or one of 16 cryptographic keys.

What happens when you successfully present a PIN in one directory and then move to another directory with a different PIN file? Does the first PIN authentication follow you? What if you move to a directory that doesn't have a PIN file, but there is a file in this new directory with an operation that is protected by a PIN? Does your old PIN work? The same questions obviously apply to cryptographic key authentication.

Smart cards vary as to their authentication-state maintenance policies and the ISO standards are mute on the subject. What we describe here is the policy of your Multiflex card. Do not assume that all cards behave the same way.

Authentication state in the Multiflex card is carried in three 2-byte variables:

- PIN authentication directory—The fileId of the directory that contains the PIN file of the last PIN to be authenticated.

- External authentication directory—The fileId of the directory that contains the external authentication file, which in turn contains the last cryptographic key to be authenticated.

- External authentication keys—16 bits with the interpretation that if bit 1 is set, then key 1 in the current external authentication file has been authenticated.

When an access-controlled operation is about to take place, the operating system simply checks these variables. If PIN authentication is required and the PIN authentication directory variable is non-null, then the operation is allowed to proceed. Otherwise, it fails. If a cryptographic key authentication is required, the external authentication directory and external authentication key variables are checked. If the first is non-null and the bit corresponding to the required key in the second is set, then the operation is allowed to proceed. Otherwise, it fails.

Moving from one elementary file to another within a directory does not change the setting of the authentication state variables since all elementary files in a directory use the same PIN and external authentication files for their access controls. Selecting a new directory does change the authentication state. Figure 5.1 shows the procedure for the PIN authentication directory variable.

The procedure for the external authentication directory and external authentication keys variables is exactly the same except the external authorization directory and externalization authorization key's state variables are cleared rather than the PIN authentication directory state variable.

The bottom line of this procedure is that authorizations don't follow you as you hop from one directory to another on the card unless the files in the new directory use the same authorization files as the files in the old directory. This seems perfectly reasonable. What PIN and external authentication files is a particular file governed by? If there is a PIN or an external authentication file in the same directory as the file, that's the one the file uses. If not, search from the directory back toward the master file and use the first PIN or external authorization file—depending on what you are looking for—that you find.

In the frequent case that you move into a subdirectory of the current directory, your authorizations stay with you unless there is a new authorization file—PIN or external—in the subdirectory, in which case the corresponding authorizations are cleared and you have to show possession of the new key(s) in order to regain them.

From the point of view of a smart card system designer, if you put a key file in the directory of a particular application, that key file defines the keys needed to access the files for that directory and all its subdirectories. If you don't provide key files in an application's directory, the key files in the directories above the application define the keys needed to access the application's files. In other words, your Multiflex card implements *hierarchical access control*.

In a typical scenario, there will be one PIN file in the master file that establishes cardholder authentication and access privileges throughout the card. Then there are external authentication files with each application that govern who can read data gathered by the application and the parameters of running the application.

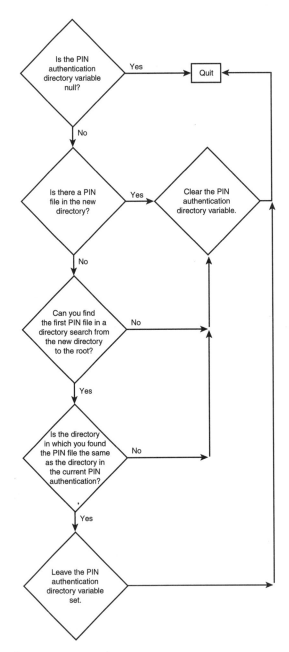

Figure 5.1. *PIN authentication procedure.*

Tracking EEPROM Usage

Nonvolatile memory (NVM) is a precious resource on a smart card and you will find yourself keeping track of every byte. Each file in NVM takes up some extra administrative bytes besides the bytes you actually use. These overhead bytes describe the file, including its size, its type, and its access conditions. Furthermore, each record in a record file also contains some overhead "green bytes." Table 5.15 gives the size of these per-file and per-record overheads:

Table 5.15. EEPROM usage in Schlumberger's 3K Multiflex card.

File Type	File Type Byte	File Header Overhead	Per Record Overhead
Directory file	38_{16}	24 bytes	
Transparent file	01_{16}	16 bytes	
Record file with fixed-length records	02_{16}	16 bytes	4 bytes
Record file with variable-length records	04_{16}	16 bytes	4 bytes
Cyclic file	06_{16}	16 bytes	4 bytes

The information we received back from selecting the master file said that there were 2,832 unused bytes on what is advertised as a 3,008 byte Multiflex card. Where are the 176 missing bytes? The numbers in Table 5.16 and a little extra knowledge of the Multiflex operating system give a full accounting of the missing 176 bytes.

Table 5.16 Overhead bytes in the virgin 3k Multiflex card.

Description	Overhead
Operating system	26 bytes
Header information for master file (3F00)	24 bytes
Header information for serial number file (0002)	16 bytes
Contents of serial number file (0002)	8 bytes
Header information for protection key file (0011)	16 bytes
Contents of transport key file	38 bytes
Flags, testing, random number generator, ATR	48 bytes
Total	176 bytes

Summary

This chapter is by no means a substitute for complete documentation of the 3K Multiflex card. Complete documentation would cover every bit of every command, in every file and every card state, condition, and response. Here we have merely tried to give you a first-hand sense for the mind-set of a smart card by considering in detail a few selected features of one real one. Above all, observe that a smart card is very suspicious. Its basic attitude is to trust no one and to always assume that there are hostile forces out there trying to trick it.

Your task as a smart card application programmer is to mold this suspicious attitude to your particular situation; to allow the card to trust just enough to get the job done but no more. Initially, you will probably err on the side of trusting too much; of not putting in enough protection; of not tailoring the access conditions carefully enough. It is a lot easier to think about how to get a job done than it is to think about how to prevent somebody from not doing the job correctly. After you've designed your smart card application, it will be a good idea to give it to a cantankerous friend and challenge them to break it.

PART II

SMART CARD SOFTWARE DEVELOPMENT

CHAPTER 6

SMART CARD SOFTWARE DEVELOPMENT TOOLS

There are no stand-alone smart card programs. Smart card software, by its very nature, is distributed. Some parts of a smart card program exist on the smart card itself and some parts are on the terminal accessing the card. Add to this "smart" smart card readers which provide value-added interfaces to smart cards as well as network computers which may also have to be consulted during a smart card transaction. You can see that building and debugging smart card software is as much a matter of getting the bugs out of the new code as it is ensuring the new code meshes efficiently and correctly with existing code.

Even though you may be developing software for only one site of many in a typical smart card application, you may well find that you will want to have tools that apply to other sites when you need to track the consequences of your software through the entire system. It is also possible that you'll need to update or extend these other sites to accommodate your new smart card application.

Another challenge to smart card programming is the nature of the smart card itself. A smart card is carefully engineered to resist efforts to see inside it. It can't tell the difference between a hacker trying to make it divulge its secrets and a legitimate programmer just trying to find out why it isn't behaving the way he thinks it should. A smart card is constructed to not let anybody see inside; this provides another challenge to the smart card programmer.

This chapter discusses the various tools that are available to help programmers develop smart card programs. The tools are divided into eight categories, which are the first eight sections of the chapter. Within each section is a general discussion of the category, and a table of all known tools in the category. The first section is about tools for host software, the second section is about tools for card software, and the final section describes a collection of miscellaneous other tools. The current contact information for each of the tools can be found at the book's web site (www.scdk.com).

Tools for Host Software Development

For the most part you will be developing application software that *uses* smart cards as opposed to software that *runs* on smart cards. From the point-of-view of an application running on the host, a smart card is just another peripheral, albeit with some unique properties. Today, smart card peripherals are not as well-integrated into programming languages and development environments as hard disks or graphic displays. While this situation is changing quickly due primarily to Microsoft's PC/SC efforts (see Chapter 7, "Reader-Side Application Programming Interfaces"), working with smart cards from your host application can be awkward and inefficient because you have to use stand-alone tools and proprietary, device-specific function libraries. Nonetheless, there are some good host software development tools available and even though they aren't fully integrated into your development environment, they are certainly better than hex debuggers.

Smart Card Editors

The discussion begins with the Swiss Army knife of smart card software development: the combination smart card browser, editor, and formatter. In the interest of brevity, we will refer to these tools as simply *smart card editors*.

A *smart card editor* is an indispensable software development aid and is part of every smart card programmer's toolkit. The most basic editors simply let you send numeric commands to the smart card and display the numeric results returned from each command. While this may seem primitive, when you are debugging your own card-side software or trying to discover what the documentation for a card's command means, this may be both the best and the only way of exploring a card's behavior.

More elaborate editors let you pick text descriptions of the commands, provide you with helpful prompts for the arguments of the commands, and translate and display the results in more compelling ways than strings of hexadecimal digits. For example, the editor may have a table of the error codes returned by the card and display a text description of the error rather than just its numerical code. An editor might also know how to interpret state and transaction information returned by the card and be able to translate status codes into text and display them in labeled windows.

Finally, some editors support a macro language capability so that you can make a file out of a series of commands and apply this canned script quickly to a bunch of cards. In this mode, a smart card editor can be used for small-batch card personalization.

In spite of international standards, there is wide variation in the commands and command responses of cards from different manufacturers and even among cards from the same manufacturer. No smart card editor could be expected to know all the existing cards let alone keep track of new cards. Therefore, the most handy smart card editors read the description of the smart card (and in some cases the smart card reader) from a text file external to the editor itself. Such a smart card description language capability lets you add new cards to the capabilities of your editor without having to wait for the editor's author to get around to including the card in the suite of cards the editor understands. Unless you are always going to be working with the same smart card, you will want to be able to customize your smart card editor to handle new smart cards.

While there are interoperability standards for the commands supported by smart cards, primarily ISO 7816-4, a particular card is not obliged to implement all commands and may, in addition, implement commands that are unique to the card or to the company producing the card. As a result, all but the most primitive formatters include a way of describing cards and commands they support. (See Table 6.1.)

Table 6.1. Smart card browsing, editing, and formatting programs.

Product	Company	Telephone	WWW	Email
AviSIM Toolbox	AU-Systems	+468 726-7500	www.ausys.com	ahg@ausys.se
Bsmart	3-GI	+1 757 564-1834	www.3gi.com	kit@3gi.com
CardEdit	Aladdin Systems	+972 3 636-2222	www.aks.com	sales@aks.com
EZ Formatter	Strategic Analysis	+1 703 527-5410	www.sainc.com	info@sainc.com
WinPractis	Schlumberger	+1 609 234-8000	www.slb.com/et	smartcards@slb.com

EZ Formatter by Strategic Analysis, Inc.

EZ Formatter is a Windows 95 program that can browse, edit, and format smart cards in serial (COM) and PCMCIA smart card readers. Buttons on the toolbar let you turn the reader's power on, reset the card, present authentication keys, change the program configuration, select and delete files on the card, and examine the log of the bytes that have been sent to and received from the card. It also contains a status message box that shows various status codes returned by the card. Figure 6.1 shows the main toolbar of EZ Formatter.

Figure 6.1. *The EZ Formatter toolbar.*

Another status display, at the bottom of the program's window, displays in both hexadecimal digits and in descriptive text the data and SW1 and SW2 condition codes returned by the card. Figure 6.2 shows the status display of EZ Formatter.

Figure 6.2. *The status display window of EZ Formatter.*

Since smart cards maintain state, it is often convenient to review the entire sequence of commands that have been sent to a smart card along with the responses of the card to these commands. The log facility of EZ Formatter always contains a complete record of the traffic between the card and the host. Figure 6.3 shows the log display.

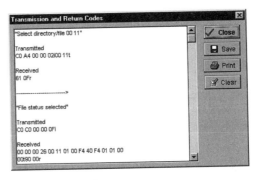

Figure 6.3. *The EZ Formatter log window.*

In a typical editing session, you will be creating and deleting files from the card, so let's take a look at the authorizations on the master file, 3F00, by selecting this directory and clicking the View Attributes toolbar button. Figure 6.4 shows the file attributes display of the Multiflex 3K master file.

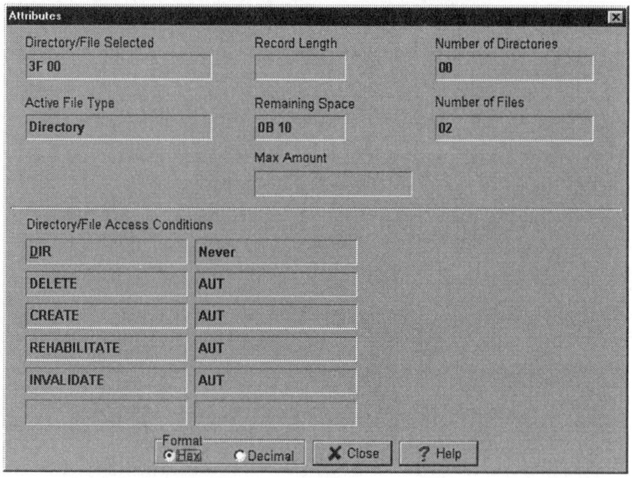

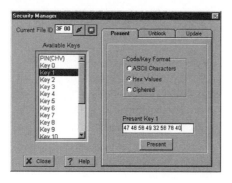

Figure 6.4. *The EZ Formatter Attributes window.*

You have to present the external authentication key in order to make any changes in the master file, so select the Security Manager toolbar button and enter Key 1, the Multiflex external authentication key. Figure 6.5 shows this interaction.

Figure 6.5. *The EZ Formatter Security Manager window.*

The latest version of EZ Formatter contains seven interaction windows that are accessed using tabs. Four of these modules—Builder, Editor, Card Controls, and Advanced Controls—can be used with any smart card. One module, Purse, is appropriate primarily for cards containing electronic purses. The remaining two, Encryption (Cryptoflex) and LoadSolo (Cyberflex), are specific to particular smart cards from Schlumberger. Figure 6.6 shows the Builder module interface.

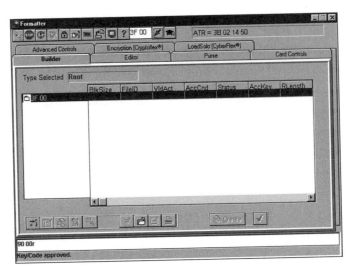

Figure 6.6. *The EZ Formatter builder window.*

The Builder module is used to create and delete new files on a card. To create a new file of 20 16-byte records, you would first describe this file to the Builder module using the window in Figure 6.7.

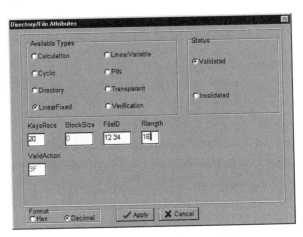

Figure 6.7. *The EZ Formatter Directory/File Attributes window.*

After clicking Apply, choose the access conditions for this new file named 1234 using the window, shown in Figure 6.8.

After you click Apply in this window, the new file shows up back in the Builder window, shown in Figure 6.9.

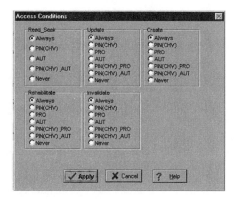

Figure 6.8. *The EZ Formatter Access Conditions window.*

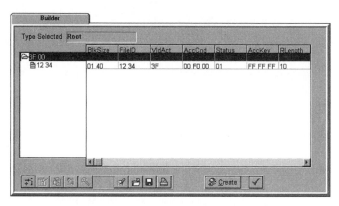

Figure 6.9. *The EZ Formatter Builder interface window.*

You have described the new file, but the file hasn't been created on the smart card yet. You can add and delete files until you are satisfied with the structure and then click Create to create all the files you've defined on the card. Furthermore, you can save this structure in a template so that it can be quickly applied to other cards.

Once the file is created, you can use the editor to read, write, and update information in the file. The file Editor window is shown in Figure 6.10.

Figure 6.11 shows the results of writing data into and reading data out of our new file.

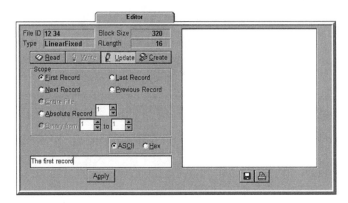

Figure 6.10. *The EZ Formatter Editor window.*

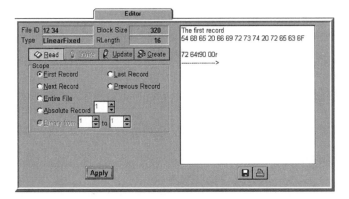

Figure 6.11. *The EZ Formatter Editor window data display.*

Finally, the Advanced Controls module can be used to execute any command the card supports. You can also use it to send arbitrary byte sequences to the card. EZ Formatter includes a simple hex editor for those situations where you have to get down to the bits. Figure 6.12 shows the Advanced Control window.

If you are going to use one of the card commands that EZ Formatter knows (as opposed to sending your own byte sequence), the command, as it will be sent to the card, is displayed along with the arguments it takes. When you click the Build Command button, an input window for each argument is displayed and a value of the argument collected. When all arguments have been supplied, you click the Process button to send the completed command to the card.

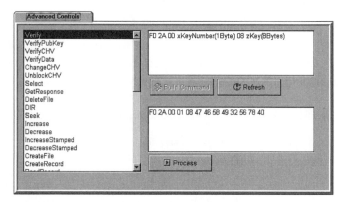

Figure 6.12. *The EZ Formatter Advanced Controls window.*

A strength of EZ Formatter is that individual smart cards and smart card readers the program can access are described in simple text files kept on your hard disk and read by the program when it starts. The language used for these descriptions is included in the EZ Formatter User's Guide. As a result, you can update the program to work with new cards and new readers that weren't in existence when you bought the program. You do this by creating a new description text file that describes the new card or card reader and loading it into EZ Formatter. Here, for example, is part of the EZ Formatter description of Schlumberger's Multiflex card:

```
[ChangeCHV]
Type=0
Instruction=F0 24 00 01 10 zOldPIN(8Bytes) zNewPIN(8Bytes)
Response Length=0
Response Time=5000
Comment=Changes secret code

[DeleteFile]
Type=0
Instruction=F0 E4 00 00 02 xFileID(2Bytes)
Response Length=0
Response Time=5000
Error1=6A 82, File ID not found
```

You can see that exactly what you put in the description file is what appears in the Advanced Controls window (refer to Figure 6.12). Furthermore, the description includes the text that is displayed in the main EZ Formatter window status bar when an error code is returned from the card.

Smart Card Systems, Infrastructures, and Plug-ins

Of course, the easiest way to implement a smart card application is simply to buy the application off-the-shelf, already smart card aware. Although the situation is changing rapidly, there are unfortunately very few smart card–aware applications available today. There are, however, software products that seamlessly and invisibly add smart card capability to existing applications (see Table 6.2).

Table 6.2. Smart card systems, infrastructures, and plug-ins.

Product	Company	Telephone	WWW	Email
Avi-BoKs	AU-Systems	+468 726 7500	www.ausys.com	ahg@ausys.se
BoKS	Dynasoft	+468 725 0900	www.dynas.se	info@dynas.com
CAP-Net Plug-ins	Selenium Intl.	+1 514 933-8800	www.selenium.com	cap@selenium.com
Snare Works	Intellisoft	+1 508 635-9070	www.isoft.com	info@isoft.com
MiniCash Electronic Purse	Donpa	+358 208-330033	www.sci.fi/~donpa	raimo.kainulainen @sci.fi
Multicard Smart	Mars Electronics	+44 118 969 7700	www.meiglobal.com	matco@meiglobal.com
NetSign	Litronic	+1 714 545-6649	www.litronic.com	info@litronic.com
Prepaid Smart Card System	Amerkore	+1 703 204-0023	www.amerkore.com	amerkore@amerkore.com
Prepaid System	Amerkore	+1 703 204-0023	www.amerkore.com	amerkore@amerkore.com
SafePak	Schlumberger	+1 609 234-8000	www.slb.com/et/	info@et.slb.com
SmartGate	V-ONE	+1 301 838-8900	www.v-one.com	sales@v-one.com
Web Browser Plug-ins	Innovonics	+1 602 516-1341	www.innovonics.com	questions@innovonics.com

Adding Smart Cards to Network Applications

Both Snare Works and SmartGate are software packages that can add smart card security to any TCP/IP infrastructure and thus to network applications. Both packages consist of a small client-side module and a larger server-side module. The client-side module invites the user to enter a PIN, then interrogates the user's smart card using this PIN to determine user identity. The client-side module then communicates the identity to the server-side module along with a request for service from the server. This design improves network security because the user needs two things—a PIN and a card—rather than just one—a password—while at the same time increasing user convenience because one card and PIN can be used to access many services.

The administrator of the server adds identity- or group-specific access controls to the server's contents. When the user asks for an item from the server, the server-side module checks the user's identity against the access control list associated with the item. If the user has been authorized to access the item, the server-side module lets the request go through to the server which returns the item. If the user is not authorized to access the requested item, a request rejection notice is returned to the user.

Intellisoft's Snare Works, as its name implies, uses OSF's Distributed Computing Environment (DCE) for directory services and secure data transport. Snare Works comes with a graphical content administration program that makes it easy to specify rules for those who can access what in the server. These rules can be very simple, such as "Sally Green can read the Delphi Project Requirements document," to very complex, such as "Senior managers in the water division can access the cover page of the electric division's monthly reports." Intellisoft calls this rule system the *Adaptive Security Framework*. Each TCP/IP packet arriving at the server is examined and processed on a protocol-by-protocol basis. For example, the contents of a World Wide Web server could be smart card protected, whereas email could flow uninterrupted. It is relatively easy to add new protocol filters, called Protocol Support Modules, for proprietary TCP/IP protocols to the system.

Intellisoft is going to expand the role of smart cards in its architecture. It will be adding to the smart card PKCS#11 and Microsoft CAPI interfaces online card personalization, multiple credentials per card, on-card key generation, and strong signing encryption.

V-ONE's SmartGate offers a more coarse-grain access control than Snare. V-ONE's rules are defined in terms of user connectivity rather than server content. Thus, for example, a user could be permitted access to email on a server but not to FTP or Telnet services on that server. SmartGate currently uses a proprietary directory and has plans to support existing industry-standard directories using the Lightweight Directory Access Protocol (LDAP).

Both systems provide for encrypted data transfer between client and server, and can keep extensive logs of traffic as a side benefit. DCE/Snare also provides, free of charge, encrypted data transfers between clients.

Smart Card Software Development Kits and Application Programming Interfaces

Most smart card applications consist of custom host software running against an off-the-shelf smart card or against a smart card standard. A growing number of smart card software development kits (SDKs) and application programming interfaces (APIs) make this an easy task (see Table 6.3). Some of these are card or card-reader specific, but opening of the smart card application development marketplace is beginning to force interoperability standards on the makers of smart card system components, so that this is becoming less rather than more of a problem.

Table 6.3. Smart card software development kits and application programming interfaces.

Product	Company	Telephone	WWW	Email
CryptOS	Litronic	+1 714 545-6649	www.litronic.com	info@litronic.com
Ecash SDK	Digicash	+31 20 592-9999	www.digicash.com	info@digicash.nl
EZ Component	Strategic Analysis	+1 703 527-5410	www.sainc.com	radclm@sainc.com
IBM Smart Card	IBM	+44 171 202 3743	www.chipcard.ibm.com	alasdair_turner_Toolkit@uk.ibm.com
ICDKT1	AmeriSys	+1 514 620-8522	www.login.net/amerisys/	info@amerisys.com
IC-XCard	HealthData Resources	+1 512 306-1926	www.hdata.com	sales@hdata.com
KapschCard Development Tools	Kapsch	+431 811 110	www.kapsch.co.at	zeppelza@kapsch.co.at
MASDAK	Integrated Technologies	+1 612 941-3605	www.itaincorp.com	sales@itaincorp.com
OSCAR Application Generator	Oberthur Smart Cards	+1 310 884-7900	www.kirkplastic.com	david.ankri@wanadoo.fr
PC/SC	Microsoft	+1 512 331 3128	www.smartcardsys.com	pcsc@slb.com
SignaSURE	DataKey	+1 612 890-6850	www.datakey.com	sales@datakey.com
SM/SW/1.1	GIS	+44 1223 462200	www.gis.co.uk	christopher@gisltd.demon.co.uk
Smart Card ADK	Amerkore	+1 703 204-0023	www.amerkore.com	amerkore@amerkore.com
SmartStart	American Magnetics	+1 213 775-8651	www.magstripe.com	webmaster@magstripe.com

EZ Component

EZ Component from Strategic Analysis is a plug-in for Borland's Delphi (Pascal) programming environment. It is especially useful because it externalizes the interface descriptions of smart cards and smart card readers so that you can relatively easily add a new card or a new reader to the system. Further, the external descriptions are what you program against when building a smart card application. So, in a sense, you can build your own smart card and smart card reader API with this package.

Here's a simple example of how you can use this external description capability to connect a smart card application that has been written independently of any particular smart card to a specific smart card at hand. Consider the command `Verify` used to present a key value to the card. The application would use this command to gain authorization to do certain card operations such as creating a new file. In the card description file for a specific smart card, the `Verify` command might have the following representation:

```
[Verify]
Instruction=F0 2A 00 xKeyNumber 08 zKeyValuecode
```

where `xKeyNumber` represents the index, in hex, of the key value to be presented and `zKeyValue` represents the actual key value in ASCII. To check whether key 1 is the word `mizpixie`, the application would include the following line of code:

```
ExecuteCommand(['Verify', '1', 'mizpixie'])
```

When this function is executed, EZ Component uses the expansion of the Verify command found in the card's description file to send the following byte stream to the card:

```
F0 2A 00 01 08 6D 69 7A 70 69 78 69 65
```

Using Borland Delphi, you could build this tiny application by putting the EZ Component and Verify button on a form as shown in Figure 6.13.

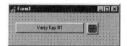

Figure 6.13. *A very simple smart card application program window.*

Then you attach the code shown in Figure 6.14 to the Verify button.

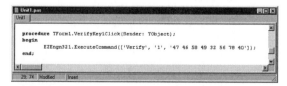

Figure 6.14. *The Borland Delphi code control screen.*

Besides the initialization function, there is only one function on the EZ Component API: `ExecuteCommand`. This function simply builds byte streams according to the recipe's description file using the arguments provided in the function call and sends the streams to the card.

EZ Component comes with description files for a number of popular smart card readers and smart cards, so you can get going right away. You have some good examples to use in extending the system when necessary. The package also comes with the source code for a simple smart card browser written using EZ Component, which can be a starting point for your application.

Smart Card Reader Interfaces

Smart card readers connect a smart card to a personal computer or a workstation. The connection can be through a serial port, a parallel port, a PCMCIA port, a keyboard port or even the floppy disk slot on the computer. A smart card reader provides power and clock to the smart card and opens up a communication channel between application software on the computer and the operating system on the smart card. Almost all smart card readers are actually reader/writers in that they allow your application to write on the card as well as to read it.

Some smart card readers are called pass-along readers because they just pass along to the smart card the byte sequences that are provided by the host application and pass back to the host application bytes that come out of the smart card. Other readers support their own command set so that your application has to be prepared to communicate with the reader as well as with the card. These smart smart card readers typically come with a software library that is intended to make communicating with the reader as well as the card inserted into the reader easy.

As you will see in the next descriptions of smart card reader APIs, there is no general consensus today of how much functionality belongs in a smart card reader. Some readers, such as the pass-along Litronic 210, provide no added semantics to the communication between the host and the card. Others, such as the Oki Value Checker Plus, are computers in their own right complete with keyboard and display. It is not difficult to imagine smart card applications for which one or the other of these readers would be appropriate.

As smart cards become integrated into operating systems, smart card reader interfaces will go away. Just as your application program doesn't know or care about the manufacturer of the hard disk controller in your PC, it will not need to know or care about the manufacturer of the smart card reader.

Nevertheless, the real situation today is that your application may have to be aware of the smart card reader with which it is communicating. What is worse is that there are smart card readers on the market that only handle particular cards. Unless you have a compelling reason to do so, you should avoid reader-specific SDKs and APIs and use general-purpose smart card access interfaces such as PC/SC (see Table 6.4).

Table 6.4. Smart card reader interfaces.

Product	Company	Telephone	WWW	Email
ACR Reader	AND	+31 10 4367100	www.and.nl	marissa@and.nl
ACR20	Advanced Card Systems	+852 279-54877	www.acs.com.hk	info@acs.com.hk
ASESoft	Aladdin Knowledge Systems	+972 3 636 2222	www.aks.com	sales@aks.com
E-Key Readers	ADC Technologies	+65 743 8088	www.adc.com.sg	msg@adc.com.sg
G80-1500 Keyboard	Cherry Electrical Products	+1 847 662-9200	www.cherrycorp webmaster.com	@cherrycorp.com
SmarTLP	Bull CP8		www.cp8.bull.net	
Smart Card Reader Library	Protekila	+90 212 261 01 63	www.protekila .com.tr	info@protekila.com.tr
SmartOS	SCM Microsystems	+1 408 370-4888	www.scmmicro.com	adapt@scmmicro.com
SmartPort	Tritheim Technologies	+1 813 943-8684	www.tritheim.com	tech@tritheim.com
SwapSmart SDK	SCM Microsystems	+1 408 370-4888	www.scmmicro.com	adapt@scmmicro.com
Value Checker Plus	OKI	+1 508 460-8621	www.oap.oki.com	digitalmoney@oki.com

The APIs described in the following sections give a sense of how your application program views a reader and a card combination, and how various vendors have dealt with the problem of having two active elements—the reader and the card—at the other end of the communication line.

Aladdin ASESoft Microprocessor APIs

The Aladdin reader can handle ISO 7816 microprocessor cards as well as I^{2}C memory cards, both protected and unprotected. The Aladdin Smartcard Environment (ASE) development kit, ASESoft, provides two APIs to microprocessor cards in a number of different languages including BASIC, Pascal, and C. The first API supports file and password manipulation functions, but only for Aladdin's CC1, CP1, and CG2 microprocessor cards. The second simply passes APDUs to and catches APDUs from any ISO 7816 card.

The functions on the file and password interface are

`ASE_DRVInit`—Initializes the Aladdin reader

`ASE_CardOn`—Turns power on the card and returns its type

`ASE_CardOff`—Takes power off the card

`ASE_GetVersion`—Returns the current version of the library and the reader

`ASE_FileCreate`—Creates a file

`ASE_FileInit`—Fills a newly created file with nulls

`ASE_FileWrite`—Writes contents of a buffer into a file

`ASE_FileRead`—Reads from file into a buffer

`ASE_PassCreate`—Sets a key value and the associated number of attempts value

`ASE_PassChange`—Changes a key value

`ASE_PassPresent`—Presents a key value to the card

The 7816 interface works with any 7816 card. The functions on this simpler API are

`ASE_DRVInit`—Initializes the Aladdin reader

`ASE_7816CardOn`—Turns power on the card and returns the card type

`ASE_7816CardOff`—Takes power off the card

`ASE_GetVersion`—Returns the current version of the library and the reader

`ASE_7816Send`—Sends a command APDU to the card

`ASE_7816Receive`—Retrieves a response APDU from the card

The Aladdin SDK comes with four modest card editor utilities for memory and microprocessor cards. These let you examine and alter the contents of these cards.

Tritheim SmartPort

The Tritheim SmartPort smart card reader attaches to a serial, parallel, PCMCIA or USB port. Like Aladdin's ASESoft, the Tritheim Smart Card Development

System (SCDS) offers a high-level interface to Tritheim's own smart cards and a low-level interface to arbitrary ISO 7816 cards.

A unique feature of the Tritheim system is that the reader itself contains an internal smart card. The internal card is addressed as Card 2 while the inserted card is addressed as Card 1. This clearly sets the stage for multicard systems such as e-commerce servers which will contain many merchant cards.

The high-level interface to the Tritheim reader and card treats the smart card as a floppy disk with password-protected files. Both linear and circular files are supported in this metaphor. Records in circular files can be read and written sequentially and randomly. The interface is implemented as a Windows DLL.

The following are the functions on the high-level disk-like interface:

`Close`—Closes a file or a card

`Dir`—Returns a directory of the card's file with attributes

`Empty`—Sets a file's contents to `NULL`

`Error`—Returns the text associated with the provided error code

`First`—Resets the sequential read pointer to the beginning of the file

`Format`—Formats the card according to the provided specification

`OpenFile`—Prepares a file for reading and writing

`ReadRandom`—Reads a specified record from a record file

`ReadSequential`—Reads the next record from a record file

`ReadLast`—Reads the last record from a record file

`Status`—Resets the card and initializes it for access

`Submit`—Submits a password to the card

`Unused`—Returns the number of unused records in a record file

`Update`—Overwrites the last sequentially read record in a record file

`WriteRandom`—Writes a specified record into a record file

`WriteSequential`—Writes the next record in a record file

As an alternative to passwords, Tritheim cards and the high-level interface also supports DES-based challenge/response authentication to gain file access. The DES extension API includes the following functions:

`Cipher`—Passes a challenge to the card for encryption

`Mode`—Turns the DES extension on and off and sets the mode of ciphering

`Random`—Gets a random number challenge from the card

`ReadCipher`—Reads back the card's encryption of a challenge

`SetApplicationModule`—Selects the source of the DES encryption

`SetCipherKey`—Reads a record from a record file and sets it as the card's DES key

The low-level interface to the Tritheim reader supports any ISO 7816 card in either T=0 or T=1 mode. The following are the functions on this API:

`IFD_Get_Capabilities`—Retrieves information about the card such as ATR string and status of contacts

`IFD_Set_Capabilities`—Attempts to set the capabilities of a parameter of the card

`IFD_Power_ICC`—Turns power to the card on or off

`IFD_Get_ICC_Com_Error`—Returns the status of the communications interface to the card

`IFD_Escape`—Sends a command to the card reader

`IFD_Read_From_SC`—Returns information from the card

`IFD_Write_To_SC`—Sends a command to the card

Advanced Card Systems ACR20

The Advanced Card Systems ACR20 reader handles both memory and microprocessor cards on a serial interface. The reader will talk T=0 or T=1 to ISO 7816 microprocessor cards. Interestingly, the ACR20 takes a daughter board that will internally accept up to three additional smart cards so that the ACR20 can act as a merchant terminal.

The API for the ACR20 consists of the following functions:

`AC_Open`—Opens the communication channel to the reader

`AC_Close`—Closes the communication channel to the reader

`AC_StartSession`—Sends a reset to the card and returns the ATR sent back by the card

`AC_EndSession`—Powers off the card

`AC_Exchange_APDU`—Sends an APDU to the card and returns the response

`AC_GetInfo`—Returns status information from the currently selected reader

`AC_SetOptions`—Sets various options on the reader

There are also 18 functions on the interface that support access to memory cards.

The Protekila Smart Card Reader Library

The Protekila Smart Card Reader library is a very simple C-language DOS and Windows interface to their smart card readers. Here is the most parsimonious of all smart card reader APIs known to the authors:

```
get_atr_respond(char *atr)
send_cmd_in(char *command)
card_reader_status(char *status)
card_power_off(char *str)
```

The Cherry G80-1500 Keyboard with Smart Card Reader

Cherry provides a Windows DLL to access the smart card reader in its G80-1500 keyboard. The reader can access both memory cards and microprocessor cards. The functions of the G80-1500 DLL can be organized into roughly 11 areas:

- Initialization

  `G1500_Init`—Initializes the G80-1500 DLL

  `G1500_Resync`—Resychronizes with the smart card terminal

  `G1500_SetNotifyMessages`—Sets values for asynchronous verification

  `G1500_SetNotifyWindow`—Sets receiver for asynchronous verification

  `G1500_Exit`—Exits from the DLL

- Automatic recognition of smart cards

  `G1500_ChipcardDetect`—Provides for automatic recognition of smart card

  `G1500_LoadATRMask`—Loads the standard ATR masks

  `G1500_ClearATRMask`—Clears all ATR masks

  `G1500_DeleteATRMask`—Deletes ATR masks

  `G1500_AddATRMask`—Adds an ATR mask

- Management of the Smart Card Standard

  `G1500_SetICC`—Sets smart card standards

  `G1500_GetICC`—Reads smart card standards

- Query parameters

  `G1500_Port`—Gives the parameters of the serial port

  `G1500_UserMem`—Gives the size of user memory

  `G1500_ProtectionMem`—Gives the size of protection memory

  `G1500_SecurityMem`—Gives the size of security memory

  `G1500_SecurityType`—Gives the type of security implemented by smart card

`G1500_NumVChars`—Gives the number of digits in the PIN numbers on the smart card

`G1500_CardClass`—Gives the class of the smart card

`G1500_AddressUnit`—Gives the type of the card's address unit

`G1500_ATR`—Reads the smart card's ATR

`G1500_KeyboardId`—Gives the model and version of the keyboard

`G1500_EpromVersion`—Gives the version of the reader

`G1500_DLLVersion`—Gives the version of the DLL

- Query settings

  `G1500_SecurityStatus`—Gives the status of the security system

  `G1500_VerifyAttempts`—Gives the number of failed attempts on current key

  `G1500_ChipcardAvailable`—Gives the status of the smart card

  `G1500_GetWorkingMode`—Reads the working mode of the smart card

  `G1500_SetWorkingMode`—Sets the working mode of the smart card

- Reading of a memory card

  `G1500_ReadUMem`—Reads user memory

  `G1500_ReadPMem`—Reads the protected memory

  `G1500_ReadSMem`—Reads the security memory

  `G1500_ReadUMemA`—Reads the user memory with attributes

- Writing to the smart card

  `G1500_Write`—Writes without protection bit

  `G1500_WriteP`—Writes with protection bit

  `G1500_WriteM`—Writes with protection mask

- Deletion

  `G1500_Erase`—Erases between addresses

  `G1500_EraseArea`—Erases an entire area

- Verification

  `G1500_Verify`—Verifies the smart card

  `G1500_ChangeVerification1`—Changes the PIN code after verification

  `G1500_ChangeVerification2`—Changes the PIN code without verification

- Counters

  `G1500_ActivateCounter`—Selects a counter

  `G1500_GetCounterVal`—Reads the counter value

  `G1500_SetCounterVal`—Sets the counter value

- Error handling

  `G1500_GetError`—Returns the error status

- Help functions

  `G1500_SetActionIndex`—Sets the action index

  `G1500_GetActionIndex`—Reads the action index

  `G1500_PMemOffset`—Returns the location of protected memory

  `G1500_PMemLength`—Returns the size of protected memory

  `G1500_Flags`—Returns the configuration of the smart card

Tools for Card Software Development

If none of the off-the-shelf smart cards suit your needs, you can

- Extend an existing smart card operating system
- Modify an existing smart card operating system
- Build your very own smart card operating system

If you choose to add code to an existing operating system, you can either add machine language code specific to a particular smart card chip or, more commonly, you can add code that is chip independent and is actually executed by an interpreter that is resident on the smart card. The Java Card and Mondex's MULTOS cards are examples of this latter approach.

Since most of the chips found on smart cards are security-enhanced versions of off-the-shelf microcontrollers, almost all the tools made to develop software for these microcontrollers can be used to develop software for their tamper-resistant versions. Where there are smart card-specific instructions, often inline data blocks can be used to access them. Most of the chip manufactures maintain good resource listings of the software products that support their chips. Some of these are listed in Table 6.5.

Table 6.5. Smart card chip development tool resources.

Chip Manufacturer	Software Development Tool Resources
All—Avnet	http://www.avnet.co.nz
All—Miller Freeman	http://www.mwmedia.com/
Hitachi	http://www.halsp.hitachi.com/tech_prod/h_micon /3_h8300/dev_tools.htm#3_h8300
Motorola	http://design-net.com/csic/DEVSYS/DevTools.htm
Philips	http://www.semiconductors.philips.com /microcontrol/devtools/
Siemens	http://www.directories.mfi.com /embedded/siemens/tools.htm
Texas Instruments	http://www.avnet.co.nz/ti/micro/370_3rd.htm

Assemblers and Compilers

The basic software development tool for creating software to run on a smart card is an assembler or a compiler together with its integrated development environment. You will program in assembly language and use an assembler if you are building software to run on a particular smart card chip and are concerned with the size and execution efficiency of the code (see Table 6.6). You will use a high-level language and a compiler if you are concerned about portability across a number of different smart card chips and are concerned with maintainability and time-to-market (see Table 6.7). In this latter case, you may be compiling into the native machine instructions of the smart card chip or into the byte codes executed by an interpreter on the chip.

Table 6.6. Assemblers for smart card processors.

Company	Telephone	WWW	Email	Processor
AND Software		www.andsoft. demon.co.uk		H8
Ashling	+353 61 334466	www.ashling.com	ashling@iol.ie	68HC05
Avocet	+1 207 236-9055	www.avocetsystems .com	avocet@midcoast.com	68HC05, 80H51
Franklin Software	+1 408 296-8051	www.fsinc.com	fsinfo@fsinc.com	80C51
IAR Systems	+1 415 765 5500	www.iar.com	info@iar.com	H8, 68HC05
Introl	+1 414 273-6100	www.introl.com	info@introl.com	68HC05
Keil	+1 972 735-8052	www.keil.com	sales.us@kiel.com	80C51

Company	Telephone	WWW	Email	Processor
Motorola	+1 512 891-6179	freeware.aus. sps.mot.com	help@www.mcu. motsps.com	68HC05
TECI	+1 802 525-3458		103006.612 @compuserve.com	68HC05

Table 6.7. C compilers for smart card processors.

Company	Telephone	WWW	Email	Processor
Archimedes	+1 206 822-6300	www.archimedesinc .com	customer-service@ archimedesinc.com	68HC05
Ashling	+353 61 334466	www.ashling.com	ashling@iol.ie	
ByteCraft	+1 519 888-6911	www.bytecraft.com	info@bytecraft.com	68HC05
Ceibo	+972 99 555387	www.ceibo.com	international @ceibo.com	80C51XA
Cosmic Software	+1 617 932-2556	www.cosmic-us.com	c-tools@cosmic-us .com	68HC05
Cygnus Support	+1 415 903-1400	www.cygnus.com	info@cygnus.com	H8
Diab Data	+1 415 571-1700	www.ddi.com	info@ddi.com	RCE
Franklin Software	+1 408 296-8051	www.fsinc.com	fsinfo@fsinc.com	80C51
Hitachi	+1 800 285-1601	www.hitachi.com		H8
Hiware	+41 61 690 75 00	www.hiware.com	info@hiware. hicom.ch	68HC05, 80C51
IAR Systems	+1 415 765-5500	www.iar.com	info@iar.com	80C51
Keil	+1 972 735-8052	www.keil.com	sales.us@kiel.com	80C51
RTS	+44 1624 623841	mannet.mcb.net/rts/	rts@mannet.mcb.net	TMS-370
Sierra Systems	+1 510 339-8200		sierra@netcom.com	80C51XA
Tasking	+1 617 320-9400	www.tasking.com	sales_us@tasking .com	80C51XA

Simulators and Debuggers

If you are developing card software, particularly assembly language code, a simulator will be an invaluable tool. A *simulator* is a program that runs on your software development computer and acts like ("simulates") the microcomputer in the target smart card. After you assemble or compile your smart card program, you send the output to the simulator, which will interpret the instructions it finds and run the program to a breakpoint or let you single-step the program and watch it change the state of the simulated smart card.

You will need a simulator not only to debug your program, but also to measure its runtime behavior (see Table 6.8). You will also find it handy to do fault analysis when cards die in the field. Remember that the card can't tell you from a hacker, so watching your program run on the card itself will be difficult at best, and typically impossible.

Table 6.8. Simulators and debuggers.

Product	Company	Telephone	WWW	Email	Processors
Boardwalk	TECI	+1 802 525-3458		103006.612 @compuserve .com	68HC05
ChipView	ChipTools	+1 905 274-6244	www.chiptools.com	infocom @chiptools .com	80C51
CrossView	Tasking	+1 617 320-9400	www.tasking.com	sales_ us@tasking .com	80C51, 80C51XA
C-SPY	IAR Systems	+1 415 765-5500	www.iar.com	info @iar.com	80C51
E6805 Symbolic Host Support	ByteCraft	+1 519 888-6911	www.bytecraft.com	info @bytecraft .com	68HC05
EVM05 Interface & SIM05 Simulator	P&E Microcomputer	+1 617 353-9206	www.pemicro.com	pemicro@ pemicro.com	68HC05
HiSIM	Hitex	+1 408 298-9077	www.hitex.com	info@hitex .com	80C51
MICSIM	RTS	+44 1624 623841	mannet.mcb.net /rts/	rts@mannet .mcb.net	TMS-370
PathFinder	Ashling	+353 61 334466	www.ashling.com	ashling @iol.ie	68HC05, 80C51
ProView	Franklin Software	+1 408 296-8051	www.fsinc.com	fsinfo @fsinc.com	80C51
PseudoMax Software Simulators	PseudoCorp	+1 804 873-1947			68HC05
SimCASE Real-Time Debuggers	Archimedes	+1 206 822-6300	www.archimedesinc .com	customer -service@ archimedesinc .com	68HC05

Product	Company	Telephone	WWW	Email	Processors
Simulator/ Debugger	Avocet Systems	+1 207 236-9055	www.avocetsystems .com	avocet @midcoast .com	68HC05
Simulator/ Debugger	2500AD Software	+1 207 236-6010	www.avocetsystems	s2500ad @rmi.net .com	68HC05, 80C51
Simulators	Hiware	+41 61 690 75 00	www.hiware.com	info@hiware .hicom.ch	68HC05, 80C51
T-N-T Sim	CARDtools Systems	+1 408 559-4240	www.cardtools.com	info @cardtools .com	H8
W6805 EVMICE Debugger	Wytec Computer		www.wytec.com	wchu@wytec .com	68HC05
YADE Software Simulator	Digicash	+31 20 592-9999	www.digicash.com	info@ digicash.nl	SC21, 83C852
ZAP	Cosmic	+1 617 932-2556	www.cosmic-us.com	c-tools @cosmic-us .com	68HC05

Emulators and Testers

In developing software for a smart card, there can be a testing and debugging phase between running your smart card program in a simulator and running it on the smart card itself. In this development phase, you run your program on a chip emulator or tester.

Chip emulators and testers are special-purpose hardware rigs that contain the actual chip in the target smart card, but instrument the chip in such a way that you can still examine the internal state of the chip and single-step your program.

Unfortunately, the capabilities of emulators and testers that make them so useful to software developers also make them very useful to smart card hackers (see Table 6.9). You may encounter some difficulty in getting an emulator or tester for the smart card chip for which you are developing code.

Table 6.9. Emulators and testers.

Product	Company	Telephone	WWW	Email	Processors
CAD Tester	ADV Technologies	+33 1 41 08 33 33	worldserver.oleane.com/adv/came.htm	came@activcard.fr	
DryICE	Hi-Tech	+61 7 3354 2411	www.hitech.com.au	hitech@htsoft.com	80C51
DS-XA In-Circuit Emulator	Ceibo	+972 99 555387	www.ceibo.com	international@ceibo.com	80C51XA
Emulators	Ashling	+353 61 334466	www.ashling.com	ashling@iol.ie	80C51XA, 68HC05
ET-iC51	Emulation Technology	+1 408 982-0660	www.emulation.com	sales@pmail.emulation.com	80C51
iceMaster	Metalink	+1 602 926-0797			H8, 80C51, 6805
In-Circuit Emulator	Hitex	+1 408 298-9077	www.hitex.com	info@hitex.com	80C51
In-Circuit Emulator	Nohau	+1 408 866-1820	www.nohau.com		8051XA
In-Circuit Emulator	Lauterbach	+1 508 303 6812	www.lauterbach.com	info@lauterbach.com	H8, 8051XA
logICC	Schlumberger	+33 1 4746 5943	www.slb.com/et/	smartcards@slb.com	Any smart card
Mini-Emulators	Digicash	+31 20 592-9999	www.digicash.com	info@digicash.nl	SC21/SC27, 83C852
MMS and MMEVS	Motorola	+1 512 502-2100	www.mot-sps.com		Any Motorola chip
SmarTest	Aspects Software	+44 131 556-4897	www.aspects-sw.com	sales@aspects-sw.com	GSM SIMs
Universal Emulators	Dr. Krohn & Stiller	+49 89 610000-12	www.iceworld.de	info@iceworld.de	80C51, 6805, TMS-370

Smart Card Operating Systems

If you find that you have to create your own custom smart card, it will be a lot easier to extend and modify an existing smart card operating system than to build one from whole cloth. A number of companies license smart card operating systems that can serve as starting points for your own card (see Table 6.10).

Table 6.10. Smart card operating systems.

Product	Company	Telephone	WWW	Email	Smart Card Chip
ACOS	Advanced Card Systems	+852 2305 3633	www.acs .com.hk	info@acs .com.hk	SC24
AMOS	AMMI	+1 408 986-1122	www .ammismart cards.com	sales@ammi smartcards .com	Any
BLUE	Digicash	+31 20 592-9999	www.digicash .com	info @digicash.nl	Any
DKCCOS	Datakey	+1 612 890-6850	www.datakey .com	sales@datakey .com	Philips
HOST	Obethur	+1 310 884-7900	www .kirkplastic .com	kirk@ kirkplastic .com	SGS-Thomson ST16
IOS	Incard	+39 823 63011	194.243.170.67/ Card-eng.htm	incardbiz @mbx.idn .it	
ISOS	GIS	+44 1223 462	www.gis.co.uk	christoper @gisltd.demon .co.uk	Experimental RISC 200
MULTOS	Mondex	+44 171 726 1996	www.mondex.com	mondex@int .mondex.com	Hitachi H8
OSCAR	GIS	+44 1223 462 200	www.gis.co.uk	christoper @gisltd.demod .co.uk	OKI MSM62
PROCOS	Protekila	+90 212 261 01 63	www.protekila .com.tr	info @protekila .com.tr	Any 6805
SOLO	Schlum-berger	+1 512 331 3774	www.cyberflex .austin.et .slb.com	guthery@slb .com	Any
SPYCOS	Spyrus	+1 408 432-8180	www.spyrus .com	info@spyrus .com	Siemens

Schlumberger's SOLO smart card operating system is written in the C programming language and thus is easily ported to any smart card chip. SOLO consists of a Java Virtual Machine on top of a collection of general-purpose ISO 7816 native functions. SOLO is the operating system in Schlumberger's Cyberflex series of smart cards. Smart card chips containing SOLO can be purchased from Motorola, SGS-Thomson, Texas Instruments, and Hitachi.

The manufacturers of each smart card chip typically provide library routines for writing and erasing the non-volatile memory on their chips. They may also provide library routines for communication, cryptography, and other specialized smart card functions. Porting an operating system to another chip with the same instruction set, for example Intel 8051 or Motorola 6805, is usually straightforward.

Miscellaneous Tools

Some tools are used in stages in the life cycle of a smart card outside the software development stage—personalization tools, hardware diagnostic tools, and prototyping systems, for example—that can also be handy during the software development stage. For the sake of completeness, some of these are listed in Table 6.11.

Table 6.11. Miscellaneous tools.

Product	Company	Telephone	WWW	Email
AviSIM OTA System	AU-Systems	+468 726-7500	www.ausys.com	ahg@ausys.se
AviSIM Personalization System	AU-Systems	+468 726-7500	www.ausys.com	ahg@ausys.se
AviSIMPOS	AU-Systems	+468 726-7500	www.ausys.com	ahg@ausys.se
DataCard 150I Personalization SDK	DataCard	+1 617 988 1763	www.datacard.com	mark_iverson @datacasrd.com
The Dumb Mouse			cuba.xs4all.nl /~hip/dumbmouse.html	
Execution Analyser	Ashling	+353 61 334466	www.ashling.com	ashling@iol.ie
GePeto - SIM Personalization	Schlumberger	+33 1 4746 6869	www.slb.com/et/	louis@montrouge.ts.slb.com
HII 7 X 24 EFT Transaction System	Halcyon	+1 206 746-4361	www.halcyon.com/hii	hii@halcyon.com

Product	Company	Telephone	WWW	Email
Inverse Reader	Digicash	+31 20 592-9999	www.digicash .com	info@digicash.nl
ROM Prototyping Card	Ashling	+353 61 334466	www.ashling .com	ashling@iol.ie
Smart Card Analyzer and Manipulator			www.cypherpunks .to/scard/	
Smart Card Probes	Ashling	+353 61 334466	www.ashling .com	ashling@iol.ie
UbiQlink Personalization System	Ubiq	+1 612 912-9401	www.ubiqinc .com	dtusie@ubiqinc.com

Summary

National and international standards define smart card interfaces. ISO has specified one interface to the smart card. ETSI has defined another, SEIS a third. There are also industry-specific standards for both smart cards and smart card readers and terminals. With all of this standards activity, you'd think that seamless smart card interoperability could be assumed.

Most smart card systems have, however, been closed systems, consisting of a specific card from a card manufacturer working with a specific terminal from a terminal manufacturer. Sometimes the same company manufactured both the card and the reader. As a result, standard-specified, paper interoperability has rarely been subjected to a reality check. As we move from closed smart card systems with proprietary components to open smart card systems with inter-changeable parts, surprisingly we are discovering that conscientious people have interpreted the many standards differently and that there is a need for additional discussion and specification.

Smart card software reflects this transitory situation. Most smart card development kits are still card and reader specific. Some have externalized the card and reader descriptions so that the buyer of the kit can adapt the software to new cards and readers. There are also architectures operating system extensions such as OpenCard and PC/SC coming along that localize card and reader differences and provide the foundation to build card- and reader-independent applications.

CHAPTER 7

READER-SIDE APPLICATION PROGRAMMING INTERFACES

We discussed some very low-level reader-side application programming interfaces (APIs) to smart cards in our discussion of reader SDKs in Chapter 6, "Smart Card Software Development Tools." These APIs are primarily concerned with controlling the reader and usually view the card as simply a place to send byte strings to and receive byte strings from; it is up to the application to construct and deconstruct these byte strings. In a sense, this is the wrong way around. What an application program typically cares most about is the functionality of the card and not the card reader, just as it cares about what's on a floppy disk and not about the control signals for the floppy disk reader.

Until very recently, there were no reader-independent smart card application programming interfaces. This was true for two reasons. First, because the card reader stood between the application program and the card reader, and because card manufacturers couldn't know which reader their card would be used with, most SDKs were provided by reader manufacturers and not card manufacturers. A reader manufacturer's primary concern was, understandably, to tell people how to use its reader, not how to use the various cards that might be used with it. Furthermore, unlike cards, card readers and therefore the programming interfaces to them are not standardized. Since readers varied widely in their capabilities, each of these interfaces was unique and tended to highlight those features of the reader that made it different from its competitors. Second, because most smart card systems were closed, consisting of a particular reader used with a particular card, there was no real reason to try to define and gain consensus for a reader-independent interface to smart cards.

Is There a Windows for Smart Cards? Will There Ever Be?

For the application programmer, smart cards are like programming languages. There are those dull old standard commands and then there are those exciting new (vendor-specific) language extensions. It is very hard to resist reaching outside the standards perimeter when that vendor goodie on the other side of the line does exactly what you need done and then some. As a result, it is unlikely that smart card manufacturers will be given any marketplace incentive to compete solely on the speed, size, and cost of a standard command set in the near future.

Furthermore, smart cards are just now coming into their own as a viable general-purpose computing platform, and everybody seems to want to be the Microsoft of smart cards—probably including Microsoft. At least for the next four or five years, as smart cards get more memory and more processing power, their operating systems and command sets will sprout ever more differentiating capabilities.

This is not to say that there won't be pushback from the application building and card issuing communities for standardization. Big customers from the monolithic industries, such as financial services and telecommunications, will use their buying power to force the production of minimum-cost commodity cards. But there will be explosive growth at the edges as new applications for smart cards are discovered and needs for new capabilities are uncovered. As these new capabilities prove generally useful, they will find their way into newer versions of the standard cards.

Will there be a Windows for smart cards? One operating system for all applications provided by a single vendor? Probably not as long as the amount of memory on a smart card stays below a megabyte and candy machines aren't network devices. While space on a smart card remains at a premium, there will be economic incentives to get as many applications on the card as possible. This will mean constant pressure on the size of the operating system and probably clustering the applications into compatible families. There may be a standard card API for payment applications, a standard card API for gaming applications, and a standard card API for personal data applications, but not a standard Windows-like API for all applications.

As smart cards have come into more general use and smart card readers have become standard equipment on more and more computing platforms, this situation has started to change rapidly. Programmers want to build smart card applications that can be used with many cards and readers, and system builders don't want to be held captive by a single reader or card supplier. Perhaps the best known and most widely distributed general-purpose smart card interface is the PC/SC

API, but there are others. Table 7.1 lists some popular general-purpose—that is, application-independent—reader-side smart card APIs.

Table 7.1. General-purpose reader-side smart card APIs.

Name	Primary Vendor	URL	Email Contact
PC/SC	Microsoft	www.smartcardsys.com	charliec@microsoft.com
MULTOS	MAOSCO	www.multos.com	nick.habgood@multos.com
Open Card Framework	IBM	www.nc.com/opencard/	kai@zurich.ibm.com

As smart cards become more widely understood and used, smart card capabilities are also appearing as extensions or options on existing domain-specific application programming interfaces. Furthermore, since some application types—cryptography or credit card payment, for example—will have a particular use for the capabilities of a smart card, there are a growing number of domain-specific application program interfaces for smart cards. Tables 7.2, 7.3, and 7.4 list some domain-specific APIs for smart cards or that include smart cards.

Table 7.2. Payment system reader-side smart card APIs.

Name	Primary Vendor	URL	Email Contact
EMV'96	Europay, MasterCard, Visa	www.mastercard.com	
Visa ICC Specification	Visa	www.visa.com	
C-SET	Europay	www.eurocard mastercard.tm.fr	gdaligault@ europayfrance.fr
SET Version 2.0	Visa, MasterCard, AMEX	www.visa.com	tlewis@visa.com
Visa Open Technology Platform	Visa	www.visa.com	

Table 7.3. Cryptographic reader-side smart card APIs.

Name	Primary Vendor	URL	Email Contact
Cryptographic API	Microsoft	www.microsoft.com	boltr@ microsoft.com
PKCS #11	RSA	www.rsa.com	
DCE Personal Security Module	Open Group	www.opengroup.org	
Schlumberger Cryptoflex	Schlumberger	www.cyberflex. austin.asc.slb.com	

Table 7.4. Other domain-specific reader-side smart card APIs.

Name	Primary Vendor	URL	Email Contact
IATA 791/20.204	Int'l Air Transport Assoc.	www.iata.org	sanfordk@iata.org

The reader-side smart card API situation is by no means settled, and the standards and mind-share wars are just beginning. Furthermore, these interfaces are not mutually exclusive. One could, for example, implement the EMV'96 terminal interface within PC/SC or the PKCS #11 interface within Open Card. Due to its inclusion in the widely distributed Windows operating system, the PC/SC architecture is a solid and soon-to-be widely available foundation for including smart cards in applications.

This chapter describes some popular high-level, reader-side APIs for smart cards. These descriptions are not a substitute for the full documentation of the interface, but are intended to provide a sense of the level and orientation of the interface.

PC/SC

Led by Microsoft, a team of smart card manufacturers (including CP8 Transac, Hewlett-Packard, Schlumberger, and Siemens-Nixdorf) defined a general-purpose architecture for including smart cards in personal computer systems. The specification for this architecture was published in December 1996 and is called the Interoperability Specification for ICCs and Personal Computer Systems. It has come to be known as simply PC/SC.

A primary purpose of the PC/SC architecture and specification is to enable card reader manufacturers and smart card manufacturers to develop products independently but have the results of their efforts smoothly work together. As a result of this effort, application programmers can build smart card applications that aren't tied to particular readers or cards, and system builders can mix and match readers and cards freely. Figure 7.1 is a diagram of the Microsoft PC/SC smart card architecture.

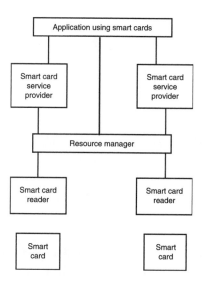

Figure 7.1. *The PC/SC architecture.*

The PC/SC architecture defines the interface between smart card readers and the resource manager so that, from the application's point of view on top of the resource manager, all smart card readers look and behave the same. A smart card reader manufacturer provides with its smart card reader hardware a PC/SC driver that connects the reader hardware to the resource manager. Thus, a smart card reader is treated by the system just like a floppy disk reader or a CD-ROM reader. The only difference is that you put a smart card into it rather than a floppy disk or a CD-ROM.

Note

Since smart card readers vary more widely and can contain more functionality than a simple floppy disk reader, there is a provision in the PC/SC architecture for the application program to communicate directly with the smart card reader in addition to communicating with the smart card that it contains. This capability is indicated by the line directly between the application and the resource manager. This interface could be of use when, for example, the smart card reader is an automatic teller machine (ATM) or a point-of-sale (POS) terminal.

The actual application programming interface seen by the smart card application is provided by the smart card service provider (SSP). This interface can be card specific, domain specific or completely general purpose. The PC/SC specification includes a description of general-purpose interface, which includes card authorization, PIN verification, file access, and cryptographic services. This interface is described in the section "The PC/SC API."

Card-specific PC/SC smart card service providers are typically written by smart card manufacturers and included with the cards themselves. Thus, for example, Gemplus provides an SSP for each of its off-the-shelf cards, as do Schlumberger and Datakey. Minimally these card-specific SSPs make each command supported by the card available to the application in an easy-to-use manner. Some also support some higher-level functionalities that can be built from the basic commands.

As smart card application domains become more well defined through various standards and specification efforts, smart card service providers that support these standards and specifications will start to appear. For example, we should soon see a SET/EMV SSP and a digital signature SSP. These domain-specific SSPs will not only support the processing and procedures that are characteristic of the domain, but they will assume cards which contain the data structures and computing capabilities that are specified for the domain. Domain-specific SSPs are prime business opportunities for third-party smart card software companies.

The PC/SC API

The PC/SC smart card API serves more as an example of how to build SSPs than it does as a commercially available and widely used smart card API. Figure 7.2 illustrates the general layout of this API.

SCARD connects to the card and maintains a context in which the other functions can operate. It has two functions, AttachByHandle and AttachByIFD, that let the application specify a card to access and includes two more functions, Detach and Reconnect, to administer this connection.

The CARDAUTH interface provides functions to enable the card to authenticate the application and the application to authenticate the card. Included on this generic interface are GetChallenge, ICC_Auth, APP_Auth, and User_Auth. GetChallenge returns a random data string from the card that is to be encrypted by the application and returned in the APP_Auth call. ICC_Auth sends a random string to the card to encrypt it and return it. Finally, User_Auth is a general interface to vendor-specific routines for user authentication.

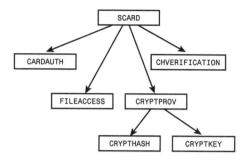

Figure 7.2. *PC/SC reference smart card API architecture.*

CHVERIFICATION is a collection of functions that connect to PIN functionality on a smart card. The functions on the interface are Verify, ChangeCode, Unblock, and ResetSecurityState. Verify presents a PIN to the card and returns success or failure. ChangeCode allows the cardholder to change the card's PIN by way of the application. Unblock lets the card's issuer unblock a PIN that has become blocked through too many unsuccessful attempts to present the PIN. Finally, ResetSecurityState causes a vendor-specific resetting of the PIN security on the card.

The FILEACCESS routines present the expected set of functions for manipulating files on the card. They are

- ChangeDir—Changes to a different directory
- GetCurrentDir—Returns the name of the current directory
- Directory—Returns a list of the files in the current directory
- GetProperties—Returns the properties of the current file
- SetProperties—Sets the properties of the current file
- GetFileCapabilities—Gets capabilities of the current file
- Open—Opens a file for access and makes it the current file
- Close—Closes the current file
- Seek—Files a data pattern in the current file
- Write—Writes data into the current file
- Read—Reads data from the current file
- Create—Creates a file in the current directory
- Delete—Deletes a file in the current directory
- Invalidate—Marks a file as unavailable
- Rehabilitate—Marks a file as available

Finally, CRYPTPROV supports some basic routines for accessing cryptographic services on a smart card. It is not the full-fledged Microsoft Cryptographic Services API (CAPI), but rather is a smart card–centric subset of CAPI that is nonetheless quite useful for adding smart card–provided cryptographic services to an application. Functions on the CRYPTPROV interfaces are

- Decrypt—Decodes an encrypted data block using a specified key
- DeriveKey—Creates keys from fixed data
- Encrypt—Encodes a data block using a specified key
- Export—Returns a key stored on the smart card
- GenKey—Creates keys from random data
- GetParm—Returns parameters being used by the routines
- GetRandom—Returns random bytes
- GetUserKey—Returns the public key
- HashData—Computes the cryptographic hash of a stream of data
- HashSessionKey—Computes the cryptographic hash of a key
- ImportKey—Provides a key to the smart card
- SetParam—Sets the parameters being used by the routines
- SignHash—Computes the signature on a hash using an asymmetric key
- VerifySignature—Verifies the signature of a hash using an asymmetric key

The Multiflex SSP

The dynamic link library (DLL) for the Multiflex SSP is included on the book's CD-ROM. This SSP can be used with the Microsoft PC/SC software to build host applications that use the Multiflex card included with the book. Instructions for obtaining and installing the PC/SC software are included on the CD-ROM.

 Note

To use the smart card included with this book, you will have to buy a smart card reader and install it on your Windows PC.

Listing 7.1 is the C header file for the Multiflex SSP API. This SSP supports both the 3K and the 8K Multiflex cards. There are some commands such as Directory, Increase Stamped, and Decrease Stamped on the 8K card that are not implemented on the 3K card due to ROM size constraints.

Listing 7.1. The C header file for the Multiflex SSP API.

```
/*++
Copyright (c); 1997  Schlumberger Electronic Transactions
Module Name:
      MultiflexSSP.h
Abstract:
      Header file for API for Multiflex DLL SSP
Author:
      Scott Guthery        1-September-1997
Environment:
      Win32
Revision History:
Notes:
      None
--*/
#ifndef _MULTIFLEXSSP_H_
#define _MULTIFLEXSSP_H_
extern HRESULT WINAPI AttachCard(WCHAR *wszReaderName);
extern HRESULT WINAPI DetachCard();
extern HRESULT WINAPI \
CreateCardFile(\
      WORD wFileId, \
      WORD wFileLength, \
      BYTE bFileType,\
      BYTE bInitialize, \
      BYTE bInitialStatus,\
      BYTE bRecordLength, \
      BYTE bNumberOfRecords,\
      LPBYTE pbAccessConditions, \
      LPBYTE pbAccessKeys);
extern HRESULT WINAPI \
CreateCardFileMAC(\
      WORD wFileId, \
      WORD wFileLength, \
      BYTE bFileType,\
      BYTE bInitialize, \
      BYTE bInitialStatus,\
      BYTE bRecordLength, \
      BYTE bNumberOfRecords,\
      LPBYTE bpAccessConditions, \
      LPBYTE bpAccessKeys,\
      LPBYTE bpCryptogram);
extern HRESULT WINAPI
CreateRecord(\
      LPBYTE bpInitialData, \
      BYTE bDataLength);
extern HRESULT WINAPI \
CreateRecordMAC(\
      LPBYTE bpInitialData, \
      BYTE bDataLength, \
```

Listing 7.1. continued

```
        LPBYTE bpCryptogram);
extern HRESULT WINAPI \
DeleteCardFile(\
        WORD wFileId);
extern HRESULT WINAPI \
DeleteCardFileMAC(\
        WORD wFileId, \
        LPBYTE bpCryptogram);
extern HRESULT WINAPI
VerifyKey(\
        BYTE bKeyNumber,\
        LPBYTE bpKey,\
        BYTE bKeyLength);
extern HRESULT WINAPI \
VerifyPIN(\
        LPBYTE bpPIN);
extern HRESULT WINAPI \
ChangePIN(\
        LPBYTE bpOldPIN,\
        LPBYTE bpNewPIN);
extern HRESULT WINAPI \
UnblockPIN(\
        LPBYTE bpUnblockPIN,\
        LPBYTE bpNewPIN);
extern HRESULT WINAPI \
GetChallenge(\
        LPBYTE bpChallenge);
extern HRESULT WINAPI \
GiveChallenge(\
        LPBYTE bpChallenge);
extern HRESULT WINAPI \
InternalAuthentication(\
        BYTE bKeyNumber,\
        LPBYTE bpChallenge);
extern HRESULT WINAPI \
ExternalAuthentication(\
        BYTE bKeyNumber,\
        LPBYTE bpChallenge);
extern HRESULT WINAPI \
Invalidate();
extern HRESULT WINAPI \
InvalidateMAC(\
        LPBYTE Cryptogram);
extern HRESULT WINAPI \
Rehabilitate();
extern HRESULT WINAPI \
RehabilitateMAC(\
        LPBYTE Cryptogram);
extern HRESULT WINAPI \
SelectFile(\
        WORD wFileId);
```

```
extern HRESULT WINAPI UpdateBinary(\
     WORD wOffset, \
     LPBYTE bpData, \
     BYTE bDataLength);
extern HRESULT WINAPI UpdateBinaryMAC(\
     WORD wOffset, \
     LPBYTE bpUpdateData, \
     BYTE bDataLength,\
     LPBYTE bpCryptogram);
extern HRESULT WINAPI UpdateRecord(\
     BYTE bRecordNumber,\
     BYTE bMode, \
     LPBYTE bpData, \
     BYTE bDataLength);
extern HRESULT WINAPI UpdateRecordMAC(\
     BYTE bRecordNumber,\
     BYTE bMode, \
     LPBYTE bpUpdateData, \
     BYTE bDataLength,\
     LPBYTE bpCryptogram);
extern HRESULT WINAPI ReadBinary(\
     LPBYTE bpData, \
     WORD wOffset, \
     BYTE bDataLength);
extern HRESULT WINAPI ReadRecord(\
     LPBYTE bpData,\
     BYTE bRecordNumber,\
     BYTE bMode,\
     BYTE bDataLength);
extern HRESULT WINAPI Seek(\
     BYTE bOffset,\
     BYTE bMode, \
     LPBYTE bpPattern, \
     BYTE bPatternLength);
extern HRESULT WINAPI \
Increase(\
     DWORD dwAmount);
extern HRESULT WINAPI \
IncreaseMAC(\
     DWORD dwAmount,\
     LPBYTE bpCryptogram);
extern HRESULT WINAPI \
IncreaseStamped(\
     DWORD dwAmount);
extern HRESULT WINAPI \
IncreaseStampedMAC(\
     DWORD dwAmount,\
     LPBYTE bpCryptogram);
extern HRESULT WINAPI \
Decrease(\
     DWORD dwAmount);
```

Listing 7.1. continued

```
extern HRESULT WINAPI \
DecreaseMAC(\
      DWORD dwAmount,\
      LPBYTE bpCryptogram);
extern HRESULT WINAPI \
DecreaseStamped(\
      DWORD dwAmount);
extern HRESULT WINAPI \
DecreaseStampedMAC(\
      DWORD dwAmount,\
      LPBYTE bpCryptogram);
extern HRESULT WINAPI \
Directory(\
      LPBYTE bpDirectory);
extern HRESULT WINAPI \
GetResponse(\
      LPBYTE bpResponse,
      BYTE bResponseLength);

extern HRESULT WINAPI GetSW(LPWORD wpSW);;
extern HRESULT WINAPI GetLr(LPDWORD dwLr);;
#endif // _MULTIFLEXSSP_H_
```

The Multiflex SSP itself simply surfaces all the Multiflex commands. The applications in Chapters 10, "The Smart Shopper Smart Card Program," and 11, "The FlexCash Card: An E-commerce Smart Card Application," use this SSP.

MULTOS

MULTOS is a general-purpose smart card operating system that evolved from the development of the Mondex e-cash card. Like the Java Card, it supports card-side programming (see Chapter 8, "Card-Side Application Programming Interfaces"), but does it in the C programming language rather than in Java. Since MULTOS is an operating system, there is no technical reason why it couldn't run Java too, or Forth or Tcl or BASIC, for that matter.

The simplest MULTOS API is one that surfaces the native MULTOS commands, but this is not a very interesting API. The power of being able to do card-side programming is that you can build your own command set and move functionality between the host and the card much more readily and with greater ease. Since the strength of MULTOS is in its card-side programming, we will discuss it further in Chapter 8.

The application program license fee is fairly small. If you want a copy, contact Nick Habgood at nick.habgood@mondex.com and become a licensed MULTOS application developer.

The Open Card Framework

The Open Card Framework was produced by IBM in conjunction with Netscape, NCI, and Sun Microsystems around the same time as the PC/SC architecture. It is somewhat more complex than the PC/SC architecture and was designed with the use of a smart card in a network computer in mind. Since network computers have been slow in coming, the framework has suffered from a lack of actual use.

One implementation of the Open Card Framework, the IBM Java Smart Card API, was created for JavaOS–based network computers, but there are very few smart card readers available for this system.

Since the PC/SC architecture is a platform-independent design, you might think that it would make sense to implement PC/SC in network computers as well as Windows PCs, rather than to define a wholly different smart card architecture for these computers.

Given that many of the same companies participate in both the PC/SC Workgroup and the Open Card Framework Group, there are hopeful signs that PC/SC and Open Card will be harmonized and that the attractive features of Open Card will be folded into and strengthen PC/SC.

ICC Specification for Payment Systems (EMV'96)

The ICC Specification for Payment Systems, popularly known as EMV'96, is produced by three bank card associations: Europay (www.europay.com), MasterCard (www.mastercard.com), and Visa (www.visa.com). (The abbreviation *EMV* is derived from Europay, MasterCard, and Visa.) It has undergone a number of transformations since it was first published in 1993. The current version, version 3.0, was published June 30, 1996, hence EMV'96. The specification is available in its entirety from the Web sites of all three associations.

Initially, the specification was simply intended to describe a closed proprietary system for handling credit and debit smart cards so that vendors could build interoperable and price-competitive hardware that banks and merchants could purchase to implement the system. Electronic commerce burst on the scene and MasterCard and Visa began to develop the SET (Secure Electronic Transactions) specification to handle credit and debit cards on the Internet; the EMV specification is being evolved by its sponsors into a specification for the SET smart card.

The EMV specification comes in three parts, which define

- The smart card
- The terminal
- How the application is expected to interact with the card

Thus, EMV is more than just an API. It is a specification for smart card transaction processing. It is also unique in that it is the first detailed specification to describe a smart card that contains more than one application; that is, a multiapplication smart card. The applications on which it focuses are payment applications, but it would not be difficult to use the EMV specification as a starting point for the description of how an application should interact with a universal multiapplication smart card. If multiapplication cards are to be widely used, such a protocol will be needed and, to the extent that they want to be the owners of the consumers' smart card platform, this may be what the bank card associations have in mind in the long run. Figure 7.3 shows the overall architecture of an EMV smart card.

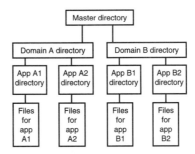

Figure 7.3. *EMV card layout.*

A *domain* is a collection of similar applications. The similarity is based on both the type of data and transaction protocols used by the applications. EMV'96, for example, defines the domain of payment system applications. IATA 791, discussed in the section "IATA 791/20.204," defines the domain of air travel applications. The directory for each domain has a unique name and a terminal picks a domain by simply resetting the card and selecting the domain's directory in the master directory. The EMV payment system directory is 1PAY.SYS.DDF01. The IATA air travel domain directory is FLY.SYS.DDF01.

The general steps in an EMV payment system transaction are as follows:

1. Application selection

2. Risk management

3. Transaction processing

4. Card updating (optional)

The application selection and card updating steps are not particular to the payment system domain and could be incorporated in any domain. The risk management and transaction processing steps are definitely specific to EMV's payment system domain. Each application in the EMV payment system domain is a card from an issuer. For example, one application might be a One Bank credit card and another application might be a Second Bank debit card.

Each entry in the payment system directory file, `1PAY.SYS.DDF01`, describes one payment system application. In a typical situation in which the cardholder is to be asked to select a method of payment, the host program would read this file and present the cardholder with the list of payment systems that it found on the cardholder's EMV smart card. Obviously, if the host application couldn't handle a particular payment system it found on the card, it wouldn't offer it as an alternative.

After an application has been selected, the host application selects the application directory and reads a number of parameters that tailor the risk management and transaction processing steps to the policies and protocols of the particular application. One Bank may require the host to go online to its network computers to authorize all transactions, whereas Second Bank may require the host to go online only for transactions over $10. All the policy and protocol parameters are stored in application files defined by the EMV specification.

The optional card updating step is accomplished by command scripts that are provided to the terminal by the owner of each application. These scripts may be stored in the terminal, or they may be provided to the host during online interaction. For example, suppose a terminal had been instructed by the application parameters on the card to go online to authorize a transaction, and suppose further that the issuer's system determined that the card had been reported as stolen. In addition to denying the particular transaction, the issuer would send an update script or procedure to the terminal that would cause its application on the card to be blocked. In this way, the stolen card could not be used for any subsequently attempted offline transactions.

One key shortcoming of the EMV architecture is the lack of a well-defined method for sharing data (except public key certificates) within the payment system domain or across domains. The cardholder may wish to maintain one version of his home address and telephone numbers on the card in order to make updating quick and easy.

EMV Commands

EMV extends the ISO 7816 command set, and the EMV specification goes to great lengths to define the details of the specific commands that a smart card must respond to in order to be EMV compliant. The EMV smart card commands are:

- APPLICATION BLOCK (CLA=$8C_{16}$ or 84_{16}, INS=$1E_{16}$)
- APPLICATION UNBLOCK (CLA=$8C_{16}$ or 84_{16}, INS=18_{16})
- CARD BLOCK (CLA=$8C_{16}$ or 84_{16}, INS=16_{16})
- EXTERNAL AUTHENTICATE (CLA=00_{16}, INS=82_{16})
- GENERATE APPLICATION CRYPTOGRAM (CLA=80_{16}, INS=AE_{16})
- GET DATA (CLA=80_{16}, INS=CA_{16})
- GET PROCESSING OPTIONS (CLA=80_{16}, INS=$A8_{16}$)
- INTERNAL AUTHENTICATE (CLA=00_{16}, INS=88_{16})
- PIN CHANGE/UNBLOCK (CLA=$8C_{16}$ or 84_{16}, INS=24_{16})
- READ RECORD (CLA=00_{16}, INS=$B2_{16}$)
- SELECT (CLA=00_{16}, INS=$A4_{16}$)
- VERIFY (CLA=00_{16}, INS=20_{16})

Data Authentication and Digital Certificates

The EMV specification defines two methods for authenticating the data on an EMV smart card—static and dynamic. These techniques are not unique to payment system applications and could be used to authenticate important data in any application domain. Data authentication ensures that the card is authentic and not a counterfeit or "spoof" card.

Static data authentication simply checks whether unvarying data that was placed on the card when it was originally created is still valid. Dynamic data validation checks data that can change during the lifetime of the card. Both methods employ public/private key pairs for authentication.

An EMV application that supports static data authentication carries the digital certificate of the issuer's public key, along with the static data signed with the issuer's private key. The digital certificate of the issuer's public key is signed by a certificate authority whose public key is held by the terminal. The authentication of the static data is performed by the terminal as follows:

1. Retrieve issuer's public key digital certificate from the smart card.
2. Verify authenticity of the digital certificate using the certificate authority's public key.
3. Retrieve signed static data block to be authenticated from the smart card.

4. Verify the authenticity of the signed static data block using the issuer's public key.

Dynamic data authentication is a little more complicated but runs along the same lines. In order to support dynamic data authentication, the smart card carries its own private key and a digital certificate for the corresponding public key in addition to the digital certificate containing the public key of the issuer. The authentication of dynamic data is performed by the terminal as follows:

1. Retrieve the issuer's public key digital certificate from the smart card.

2. Verify the authenticity of the issuer's public key digital certificate using the public key of the certificate authority.

3. Retrieve the digital certificate of the smart card's public key from the smart card.

4. Verify the authenticity of the smart card's public key digital certificate using the public key of the certificate authority.

5. Use the INTERNAL AUTHENTICATE command to instruct the smart card to sign specific data elements using the smart card's private key.

6. Retrieve signed data elements from the smart card.

7. Verify the authenticity (and accuracy) of the signed data elements using the card's public key.

It is easy to see the beginning of the SET protocols in this simple EMV data validation protocol.

Visa Integrated Circuit Card Specification

The Visa Integrated Circuit Card (ICC) specification is Visa's extension of EMV'96 beyond credit and debit payment systems to stored value and loyalty applications. In particular, it applies EMV'96 to two Visa stored value applications, the Chip Card Payment Service (CCPS) and VisaCash.

The Visa ICC specification also covers Visa's Easy Entry smart card application. Easy Entry defines a way for a smart card to behave like a magnetic stripe card—not electrically of course, but with respect to the format and content of the data it emits. As a result, a smart card containing Easy Entry applications can be used with the existing and extensive magstripe infrastructure. It is the duty of the terminal to perform the electrical translations between the ISO 7816 interface to the smart card and the needs of the magstrip transaction processing network.

Because stored value and loyalty applications make it necessary to be able to write data to the card as well as read from it (to add cash or points to the card, for example), the CCPS specification adds a PUT DATA command (CLA=04_{16}, INS DA$_{16}$) to the basic EMV'96 commands. Furthermore, since this command which essentially mints money just might attract some hacker interest, the secure messaging and data validation capabilities of EMV'96 are also considerably strengthened in CCPS so that only the right people can increase cash value or the loyalty point totals on the card.

SET 2.0 and the Visa Open Technology Platform

As of this writing, neither the specifications for SET 2.0 nor the specifications for Visa's Open Technology Platform have been released. SET 2.0 is claimed to include smart cart support and you can imagine that it will be the next step in the evolution of EMV'96. Clearly, the stage has been set in EMV'96 and CCPS to handle SET.

The Visa Open Technology Platform (OTP) is a customization of the Java Card specification, which is a closed and highly constrained multiapplication environment. A primary concern of the Java Card and the OTP is to provide the card issuer with quality control over the applications on the card. What impact this has on cardholders and application developers remains to be seen.

C-SET

C-SET (http://www.c-set.com) stands for chip secure electronic transaction. It is the adaptation of SET to French law and its views of individual privacy. Essentially C-SET injects the French government into the SET architecture via government-controlled smart cards and smart card readers, government-issued identities and "translators" that map between C-SET and SET. The C-SET translators serve as combined French taxation authority and firewall on the Internet. Unlike SET, all encryption keys in C-SET are escrowed; that is, they were held by the government, "to avoid any encryption technique falling into the hands of customers." The French view is that guaranteeing individual privacy is the responsibility of the state.

The C-SET effort is led by Groupement des Cartes Bancaires, a French bank card association and the issuer of the Cartes Bancaires (CB) payment smart card.

The C-SET protocol is essentially identical to the SET protocol except that all transactions must go through gateways controlled by Cartes Bancaires banks, since only they have the keys needed to decrypt the messages generated by C-SET smart cards and C-SETclient software. Figure 7.4 shows how a C-SET cardholder accesses a SET merchant. Notice that the SET payment instruction must go through a Bancaires bank to get to the SET gateway.

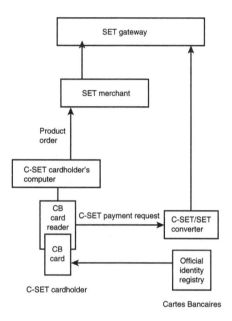

Figure 7.4. *C-SET architecture.*

The strength of C-SET is that it highlights the unique role smart cards and smart card readers can play in securing electronic commerce and establishing network identity. A smart card token adds a second factor—something you have in addition to the first factor, which is something you know such as a PIN or a password—to a network security architecture and is a convenient and reassuring form in which to create and store a private key. A smart card reader can check a PIN without exposing it to rogue software that might be hiding in publicly accessible and unguarded fields of computer memory.

IATA 791/20.204

While it isn't a payment system, the IATA 791/20.204 smart card API is included in the EMV'96 section because it conforms to the EMV'96 specification and shows how EMV can be adapted to application domains besides payment systems.

The IATA smart card is used by the airline industry to support Interline Electronic Ticketing (IET). Besides holding passenger information such as name, gender, and language preference, the IATA smart card holds a collection of digitally signed electronic airline tickets. As the tickets are used for passage, they are invalidated and boarding information is placed on the smart card.

IATA 719/20.204 uses the SELECT, GET PROCESSING OPTIONS, and READ RECORD commands of EMV'96 and adds its own WRITE RECORD command in order to write the boarding data to the card.

Cryptographic Smart Cards

The only safe place to store your private key is on something you can take with you and that won't reveal it under any circumstances. Portability means there is only one copy of your private key. If it is in something you can pop in your pocket, you can make it difficult to get the key out (tamper resistant) and it would be obvious if somebody tried to take it from you (tamper evident). A smart card is a leading candidate for private key storage.

Since the key isn't coming out of its storage place, all the operations you want to perform with it must be performed inside the card. Thus, the functionality on the card simply replaces low-level functionality in existing cryptographic packages. Rather than performing the processing in a software routine on the host, the routine is moved to the cryptographic smart card and the data shipped to the card for processing when the routine is called by higher-level routines. This is exactly the tack taken to integrate smart cards containing private keys to the Microsoft cryptographic application programming interface (CAPI) and the RSA PKCS #11 Cryptoki application programming interface. The cryptographic application using these APIs can't tell if the cryptographic functions are being performed on the host computer or in a smart card, thus there is no direct connection between the application and an API for the smart card, per se.

Cryptographic Smart Card Commands

There are a number of cryptographic smart cards on the market which can be seamlessly connected to higher level interfaces, as discussed above, but which also sport their own application programming interfaces. The Schlumberger

Cryptoflex card, the Gemplus GPK2000 and GPK4000 cards, the Datakey SignaSure cards, and the Giesecke & Devrient STARCOS family of smart cards are examples of cryptographic smart cards.

In addition to the usual ISO 7816 commands, such as SELECT FILE and READ RECORD, these cards also support commands specific to their cryptographic use. The Giesecke & Devrient STARCOS cards, for example, include the following commands:

- CRYPT
- EXCHANGE CHALLENGE
- KEY STATUS
- COMPUTE SIGNATURE
- VERIFY SIGNATURE
- PUT HASH
- SET HASH
- COMPUTE HASH
- MANAGE SECURITY ENVIRONMENT
- ASYMETRIC INTERNAL AUTHENTICATE

The Schlumberger Cryptoflex cards include the following cryptographic commands, which also appear on the PC/SC SSP API for Cryptoflex:

- RSA INTERNAL AUTHENTICATE
- FULL DES INTERNAL AUTHENTICATION
- LOAD CERTIFICATE
- VERIFY PUBLIC KEY
- VERIFY DATA
- KEY GENERATION
- UPDATE KEY ENCIPHERED

The DCE Personal Security Module API

The primary use of the DCE Personal Security Module (PSM) is to enable public key sign-on to DCE-enabled systems. Public key sign-on is more secure than the traditional password sign-on because it is two-factor (what you know and what you have) and because private keys are harder to guess than passwords. They are also less likely to be written down on a note and stuck on the computer, if only because there is little you can do with a written private key.

Open Group RFC 68 defines an API to the DCE PSM, which consists of the following functions:

- OPEN
- CLOSE
- SIGN_DATA
- VERIFY_DATA
- ENCRYPT_DATA
- DECRYPT_DATA
- PUT_PUB_KEY
- UPDATE_PUB_KEY
- UPDATE_PASSWD

These functions on the PSM are accessed by functions on the DCE Security Login API.

Summary

You will spend most of your smart card application development effort working with one or more reader-side application programming interfaces. A critical question to ask yourself as you design your application is how much of the smart card do you want your application to use. At one extreme, you can design your application to take advantage of some unique features of a particular smart card and use an API that surfaces all of these features to your application program. At the other extreme, your application may not be aware that there is a smart card "down there" at all, when, for example, the card is hidden behind a generic cryptographic interface. Between these two extremes are applications that use cards which conform to specific standards, such as EMV or ISO 7816-4, which can be purchased from many vendors.

Since a smart card is in fact an active computer rather than just an inert storage device, you will also be faced with the question of what functionality to implement on the host side of your application, and what functionality to implement on the smart card. While it can take a noticeable amount of time to ship data to the smart card and wait for its reply, the calculations it does are done in a secure environment and the parameters that these calculations use are not exposed to compromise. Generally speaking, if it is a public calculation, do it on the host, but if there are aspects of the calculation you don't want stolen, do it on the card.

A smart card can add significantly new functionalities to almost any application, initially, perhaps, as simply a secure, portable data store, but eventually as an integral part of a truly distributed application. These new functionalities will only be realized if the unique properties of a smart card, particularly its secure computing capabilities, are thoroughly understood and carefully harnessed.

CHAPTER

8

CARD-SIDE APPLICATION PROGRAMMING INTERFACES

This chapter describes application program interfaces for card software (that is, programming interfaces for the parts of application programs written to run on the processor inside the smart card itself, as opposed to application software written to run on a terminal or a host computer connected to a smart card reader containing a smart card). A number of special considerations must be taken into account when building card software that is unfamiliar to most application programmers. These special considerations include a limited and multipart memory system, a constant concern for data security, a basic distrust of users and network peers, and the possibility that power can be removed from the computer at any moment.

Programming Considerations

It is useful to distinguish between public and supported programming interfaces designed for third-party application programmers (typically outside the organization that built the interface) and internal interfaces that are part of good software engineering. A smart card operating system, like any other operating system, is built from a collection of modules which provide services on internal system interfaces to other modules. These system interfaces may change from one operating system release to the next, depending on evolutionary demands being made on the system. Application programming interfaces, on the other hand, are much more stable system interfaces and don't change from one operating system release to the next. They are, in a sense, a guarantee of functionality from the programmers building the operating system to the programmers building applications to run on top of the operating system. Since opening smart cards up to customer-provided software is a relatively recent phenomena, it will be important to ensure that software written to run on the card is sitting on a firm foundation.

It is also useful to make the distinction between writing programs in a high-level programming language and writing programs in microcontroller-specific assembly language. When writing in a high-level language such as C, Forth, or Java, the application programmer is typically presented with a well thought-out and thoroughly integrated set of services that have been explicitly designed to work together to ease the task of writing application software. One of the design goals of a good high-level programming interface is to provide help in dealing with the special considerations of smart card programming, such as those listed in this chapter.

When card software is written in assembly language, on the other hand, one can access data anywhere in memory and call upon any available entry point. Even when these entry points are part of public interfaces, they may be from different software providers and may, in fact, place calls on each other. For example, a call on a cryptographic routine may in turn generate a call on a communication routine which in turn calls a memory management routine. It is up to the assembly language programmer to understand and abide by the rules assembly language routines must obey so that they work together successfully and don't step on each other's toes. Because space and time are at such a premium inside a smart card, smart card system software is much more tightly coupled than typical operating system software, so these programming rules are much more complex than most assembly language programmers are used to.

Counterfeit Cards

Not only is allowing third-party and end-user software onto the smart card platform a recent innovation, there are still members of the smart card community who think this is a terrible idea and shouldn't be permitted at all. Beyond simply protecting their business interests, these people worry most about the creation of counterfeit cards. A counterfeit card is a card that looks like a real card to a terminal but isn't a real card. Imagine, for example, a card that could convince a candy machine that it was a valid VisaCash or Mondex card and cause the machine to vend a candy bar without actually debiting a VisaCash or Mondex purse.

Historically, the possibility of counterfeit cards has been controlled by limiting access to information about smart cards, controlling who can put software onto a smart card, and carefully reviewing all smart card software before it is released for public use. Of late, it has become obvious that if the smart card computing platform is to flourish, it needs applications and this means thousands, not tens, of application program developers. These application programmers need access to information about smart card systems and access to the smart card processor.

To their great credit, the big issuers of smart cards don't practice "security through obscurity." Visa, for example, makes the specifications for its Chip Card Payment Service and Secure Electronic Transaction protocol available on its Web site to encourage application programmers to develop applications that use these systems. This policy has the beneficial side effect of enabling the whole world to study and comment on the security of these systems. Not only have both systems been improved through this process, but trust in the use of the systems has also been strengthened.

On the other hand, there are certainly smart card systems out there built on wholly unwarranted assumptions about the cards that will be stuck into the system's terminals. These systems can and probably will be compromised by counterfeit cards created by unethical smart card programmers. The best response is better system design, better education, and better law enforcement, instead of securing the candy machine by hiding it in a closet.

Special Considerations in Writing Card-Side Software

As we imagine new applications for a smart card, we might like to think of it as just another computing platform, but the smart card computer has a number of unique properties that you must keep in mind as you design and build these new applications. In writing reader-side software, you are constantly bumping up against the severe resource constraints of the smart card as a computing platform: low-speed communication, slow central processing unit, 8-bit data, and limited available memory. Furthermore, you must be even more aware of the security implications in the logic of your code and think as much about what your code allows that you don't want it to as much as what it allows that you do want.

Finally, the smart card programmer has to be constantly mindful of the context in which the computer carrying his or her application will be used and the nature of the systems to which it is connected. A smart card computer isn't a self-contained computing system, plugged into a reliable power supply in a warm office. Rather, it is a node in a complex transaction processing network whose user may be standing in the rain and whose power supply may be interrupted by the next bolt of lightning.

Memory

Among all the special considerations that must be dealt with in writing card-side software, the ones that will be the most constant source of pain for the workstation or PC programmer are the unusual properties of the smart card memory system. Not only does the memory system consist of three different kinds of memory, each demanding to be dealt with in its own way, but the properties of these memories can vary markedly from chip-to-chip and manufacturer-to-manufacturer.

Random Access Memory

The most difficult resource constraint for most card-side programmers to deal with is the limited amount of random access memory (RAM). Small cards have 128 bytes of RAM. Big cards have 640 bytes of RAM. Carefully note the absence of any *K*s or *M*s after these numbers. That's 128 bytes, not 128 kilobytes or megabytes.

Since it is such a limited resource, RAM is managed carefully and explicitly. Some RAM is pre-allocated to hold fixed, specific, often-used values that comprise the global state of the smart card. Some RAM is pre-allocated for specific uses, such as input and output buffers. Some RAM is set aside for general scratch use and can be used freely by code for temporary and intermediate values.

Finally, some RAM is used for communication between code modules. Whether or not these locations are in current use, and if they are in use, what they contain, depends solely on the call stack. Accessing and using this RAM requires detailed knowledge of how you got where you are. Fortunately, most smart card assemblers and linkers provide tools for effectively using this type of RAM.

The first step in writing any card-side software is getting a detailed memory map of both the chip and the operating system for which you are writing and the other applications with which your code will be run.

Nonvolatile Memory

Nonvolatile memory (NVM) on a smart card is used to store values that are expected to remain on the card from use-to-use. Account numbers, digital certificates, passwords, private keys, VisaCash, and American Airlines AAdvantage (frequent flyer) points are all examples of data that would be kept in nonvolatile memory.

NVM in smart card chips is built with a number of different technologies. Various EEPROM, FLASH, and FRAM technologies are all used. One common feature of these technologies that most application programmers aren't ready for is that it takes a noticeable amount of time—up to 10 milliseconds—to write a data value into nonvolatile memory. In fact, it is not too far off the mark to say that NVM is better thought of as a randomly addressable disk block than it is workstation-style memory. In a pinch, you can use NVM for temporary values, but you probably shouldn't make a habit of it. Keep in mind the cardholder who is standing in the rain at an ATM or at the head of a long check-out line in a grocery store.

Furthermore, NVM must be written in blocks of 4, 16, or more bytes at a time. In other words, when you want to write a byte to memory location A, you also have to erase and rewrite the bytes at locations around A; at locations A-1, A+1, and A+2, for example. Although this fact is usually hidden from the application programmer by libraries supplied by the chip manufacturer, things can go wrong and values you didn't think you touched can mysteriously change by themselves. Keep in mind that writing to one location can also trigger writing to locations nearby.

Nonvolatile memory, surprisingly, wears out. Chip manufacturers claim that their NVM is good for 10,000, 40,000, or 100,000 writes, but these figures can vary widely from actual chip to actual chip. If you have a value that is going to change a lot, it is a good idea to allocate more than one memory location for it and cycle the current value through the allocated locations. For example, write the initial value at location X, the first update at location X+1, the second update at location X+2, the third update at X+3, the fourth update back at X, and so forth. This way, you spread the wear caused by the writing of the value over multiple memory locations. Of course, this does cost you more memory both to store the update code, the values, and the index of which value is the current value.

Data stored in NVM has an advertised shelf life of 10 years. To our knowledge, no studies have been done to estimate the actual mean and standard deviation of this value. Smart cards are not bad data archives but the problem may be refreshing data that you want to last beyond 10 years.

Read-Only Memory

The card-side code you write may be stored in read-only memory (ROM). Executable code and data stored in ROM are put there when the chip is manufactured. If it's wrong, it can't be changed. This can lead to expensive programming mistakes both in time and money. It can take up to four months to get a smart card chip with a particular code body made, tested, and embedded in plastic. If your program doesn't work, you get to get in line again. Furthermore, like printing after you get production set up, it is cost-effective to produce as many chips or cards as you think you'll possibly use so you will be provided with economic incentives to produce a large number of chips. If they don't work, you get to throw out an equally large number of them.

A number of programming tricks can be employed to ameliorate this situation (for example, "bugging out" to check for corrected values or substitute code fragments stored in EEPROM at strategic points in the ROM code). Indeed, the soft mask capability of smart cards discussed in the following sections is used as much to correct ROM errors as it is to add new features to existing cards. It is also a good idea to have a number of people review ROM code and even do a public instruction-by-instruction walkthrough before the mask is sent off to the factory. In spite of all your precautions, due to its irreversible nature, ROM code is fertile ground for the application of Murphy's Law.

Soft Masks

The easiest way to add code to a smart card is to put it in nonvolatile memory. Many card manufactures make off-the-shelf smart cards that accept code after they have been manufactured. If this code is native machine instructions as opposed to byte codes for an on-card virtual machine, it is called a *soft mask* to differentiate it from the mask of native machine instructions that was sent to the chip manufacturer and hardwired into the chip's ROM. Once a soft mask has been loaded into a card, it is possible to deactivate the loading capability so that no more code can be loaded, nor any bugs in the soft mask patched.

Primarily due to the counterfeit card concerns discussed earlier, it is not particularly easy to buy cards with the soft mask capability activated. Each card manufacturer has a different policy for using the soft mask capability on its cards and these policies are themselves changing as customers demand more flexibility, control, and involvement in the features and properties of the card.

In addition, since assembly language code can directly access all of the smart card's addressable memory, including ROM, it can read the operating system and the contents of all files without restriction and quite independently of the file system's access control system. Card manufacturers regard their smart card operating systems as proprietary technology and can be expected to ensure that their code isn't exposed. Finally, few of these operating systems were designed to surface a robust, documented and stable application program interface.

For all these reasons, expect to work closely with the card's manufacturer in adding a soft mask to a smart card.

Tearing

Yet another unique feature of the smart card computing platform is that power may be turned off at any moment. In designing software for a smart card, continually ask yourself, "What state would the card be in if this were the last instruction executed, and will my code recover gracefully from this state the next time the card is used?"

When a cardholder removes the card from the card reader before all the business between the card and the terminal has been finished is called *tearing*, because the cardholder is thought of as ripping the card out of the terminal. Some readers "swallow" the entire card during use and thus prevent tearing.

 Note

The readers that swallow the entire card during use can also capture the card if they decide that there is something fishy about it. These readers are much more expensive than push/pull readers, so you should probably assume that your application could be used with a push/pull reader and therefore be exposed to tearing at some time or another.

Database programmers are familiar with the use of transaction begin, commit, and rollback to ensure that relationships between multiple values in a database are preserved in the face of power loss or other processing interruption. Some smart card operating systems implement a transaction interface to data stored in NVM, but most do not. If you need to ensure that two or more values are mutually consistent—for example, that a checking account number agrees with a debit card number, or a street address agrees with a zip code—then you will have to write your own transaction code.

What database programmers and most other programmers aren't familiar with is the non-atomic write. If a storage location contains a 4 and you overwrite it with a 6 but power is turned off in the process, then you might expect that the location will contain either a 4 or a 6 when the system comes back up. If the memory is smart card nonvolatile memory and if power is removed while nonvolatile memory is being written, you could find a value that is neither 4 nor 6 when the system returns. What's worse is that the contents of locations next to the one you were writing may also be changed to random bogus values.

Critical NVM data should never be overwritten. At least one previous value should be kept. The ISO 7816-4 specification defines—and most smart cards implement—the cyclic file type for exactly this purpose. Furthermore, it should be possible to determine, based on stored data, whether tearing occurred during the previous use of the card so that card can be restored to the consistent and correct backup state.

A technique that is used if a value must be overwritten is to perform the overwrite a bit at a time so that all the intermediate values are known and acceptable. For example, if memory location containing a number representing a value is to be incremented, then you ensure that none of the intermediate values is greater than the final value. Generally speaking, from the point of view of the card owner/issuer, it is better for cards to destroy value than to create it.

Testing and Debugging

A smart card is explicitly constructed to keep hackers at bay. It includes many hardware and software features that make looking inside a smart card difficult. Unfortunately for the card-side programmer, debugging looks surprisingly like hacking from a smart card's point of view. Thus, many of the features of a smart card that make it such an attractive, secure, personal computing platform make it at the same time very difficult to deal with as a software development platform.

In the best situations, debugging and testing of card-side software proceeds in four phases:

1. *Simulation.* During the simulation phase, you run your smart card program in its software development environment typically on a workstation or Windows PC. Calls to the smart card API are simulated, including the effects of these calls on a file system or to a communication channel. Returned values faithfully reflect the result of the call. Simulation environments allow you to single-step through your program at the source code level.

2. *Emulation.* During the emulation phase, you download your code into the actual chip that will be in the target smart card, but rather than being mounted in a card, the chip is mounted on a personality board of an in-circuit emulator. This phase is more problematic than the simulation phase. Because emulators are such excellent hacking tools, not just anybody can buy an emulator for smart card chips. A viable alternative is to use an emulator for a non–smart card version of the chip you are developing (that is, an off-the-shelf version of a chip with the same instruction set). Emulation environments let you single-step through your program at the assembly language level. The theory is that your program behaves exactly as it would in an actual smart card. The theory has not always been borne out by practice, but emulation is close to actual execution.

3. *Scripting.* You can do scripting against a simulation of your program, an emulation of your program, or an actual smart card containing your program. A *script* is simply an executable description of some task that the card is expected to perform. It is a simulation of the terminal side of a smart card application. Scripts send commands to the card and monitor the card's response, reporting an error condition if the response is not as expected. By building scripts for all the demands that will be placed on your card and running them against the card, you can make sure that your program does what it is supposed to do. Systems such as the Aspects SmarTest system (`http://www.aspects-sw.com`) are explicitly built for the purpose of testing smart card software.

4. *Integration.* The final step in card software testing and debugging is to connect all the parts and components together and run the whole system just as it would be run in live use. This is typically done in the laboratory first, then with users from your own organization (alpha test). Closed population alpha tests—all the employees in a particular building or organization, for example —are a popular way to shake out smart card systems with real users. Next, the system is released to one or two friendly customers (beta test) and finally placed in a limited field test. Even though they are full-fledged computers, the smart card form factor invites everyone handling the card to expect it to work as easily and flawlessly as a credit card.

Linking and Loading

Writing, testing, and debugging code for a smart card is only half the problem. The other half is actually getting the code in use. Unlike writing code for a desktop PC, you can't just put it on a public Web site and let everybody help themselves. Not yet, at least.

When it comes to loading code onto a smart card, we come right up against the question of whose card it is. The owners of some cards may not want your incredibly creative Hunt the Wumpus game on their card. The owners of other cards may allow your game on-board, but only after you and your code have been thoroughly checked out.

 Note

Keep in mind that the owner of the card is not necessarily the person holding or using the card. You are the cardholder of the credit cards in your wallet or purse, but if you check the fine print, you'll find that you don't own them. You do own the Multiflex smart card that comes with

continues

this book, but you probably don't or won't own the smart card provided to you by your bank. The owner of the card, not the holder of the card, will by and large have the final say as to what applications can and cannot be loaded onto the card and what conditions they must satisfy before they are loaded.

One way of controlling the code that gets onto a particular smart card is to program the loader to check for a digital signature of some sort associated with the code, and only load code that is signed by an authenticated and authorized entity. You can easily imagine, for example, that One Bank will only allow code on its cards that has been approved and signed by One Bank.

On the other hand, just as there are people who will download code from anywhere onto their personal computers, there will be individual smart card cardholders who will be willing to load code onto their smart cards with no check whatsoever. But this brings up another consideration. What if Application A decides to mess with the data for Application B? Is it B's responsibility to keep its data secure from A, is it A's responsibility not to mess with B, or is it the owner's responsibility to keep A and B from fighting? If it is B's responsibility, then can B hide its data from the card owner while it's hiding it from A?

Last but certainly not least, how does your program work with the loader to find the entry points to the API that it was written against on this particular smart card? Currently there aren't any standards or even any conventions for linking new code into an existing smart card code body.

File Design

Before writing any software for a smart card application—reader-side or card-side—you should carefully design the files that the smart card will carry to support your application. At first blush, this may seem like a trivial matter. But knowledge of your application's files will spread to both card-side software and reader-side software. If you get it wrong, there will be lots of software to change—not to mention lots of cards in the field to update or discard and replace.

Smart card file design is just as complex and perhaps more so than file design for PC or even mainframe applications. This is because the smart card file design includes careful and detailed consideration of the security policy covering the data contained in the smart card's files.

First, a number of parties with different interests and concerns may be involved—card issuer, cardholder, application provider, application sponsor, and liability carrier, to name a few. Suppose, for example, we are dealing with adding a loyalty program for a local grocer to a Mondex card issued by Second Bank. This loyalty program is written by a French software house and it is to give airline travel miles to the grocer's customers. Who gets to see what data and who gets to update what data?

Once this is sorted out, there is the issue of how you go about proving somebody is who he says he is, who gets to say what is sufficient proof and who gets to approve and revoke whose authorizations? Can the grocer prevent Second Bank from looking at the customer's loyalty point holdings? Can the airline insist that the card carry a public key certificate to authenticate code from the French software house? If so, where is the certificate kept and who can see it?

It is almost impossible to spend too much time thinking through the file system design for your smart card application.

Reader Behavior

Smart card readers are not as standardized as smart cards themselves. In fact, they aren't standardized at all. Most of the smart card systems in the field today are closed systems consisting of specific cards and specific readers that have been tested to work together. While a reader can tell what to expect from a random ISO 7816 card, a card can make few assumptions about what to expect from a random reader.

There are efforts underway to standardize readers. The EMV specification discussed in Chapter 7, "Reader-Side Application Programming Interfaces," for example, has as much to say about readers as it does about cards. When writing card software, don't assume that the reader communication channel behaves as uniformly as a serial channel on a PC or a workstation. Some readers and their software drivers have built-in notions of how a card should behave and can act in rude ways—like turning off the power—if their expectations aren't met. Things to watch out for are how quickly you have to respond to reset, how long you have to respond to messages from the terminal, and what bit patterns might cause the terminal to act in strange ways.

Reader Communication

If you are building a card-side application that has to work with existing readers and existing reader software, you will undoubtedly find yourself using the ISO 7816-4 communication protocol. The basic packet of this protocol is called an application protocol data unit (APDU) and it consists of a 5-byte header followed by a Tag-Length-Value (TLV) encoded data field. This is what most of the smart card reader applications currently expect, and if you are going to play with these terminals, you will have to talk their language.

On the other hand, if you are designing and building both the card side and the reader side of your application, it may be attractive to consider the possibility of adopting something other than ISO 7816's APDU protocol. This protocol does have a specific model of computation—namely a master/slave relationship between terminal and card—built into it and as a result is awkward and inefficient in particular situations (for example, when the card wants to control the conversation). On the other hand, designing your own communication protocol does isolate your system from the mainstream of smart card systems. This may be a bug or a feature.

Fortunately, the designers of the ISO 7816 APDU protocol were well aware of the need to provide for graceful growth and extension. The result is that you can use APDU packets but endow them with virtually any semantics you wish. First, you can define wholly new APDU packets that have your meaning and carry your data. If this doesn't provide you with enough freedom, you can use any of the 47 class byte values (D0 through FE) that have been explicitly set aside for wholly arbitrary and proprietary extensions of ISO 7816. What follows an ISO 7816 D0 header byte is completely up to you.

There is one danger lurking in the bushes should you get too creative in your communication packet design. Some readers and some software infrastructures have ISO 7816 assumptions built into them and get unhappy if what they see passing between the two halves of your application isn't sufficiently ISO 7816 by their lights. If possible, be sure to test your application with all the readers and all the infrastructures with which you intend it to work.

The Standards-Based APIs

One reason you write card-side code is to move functionality from the terminal to the card. This may be done to improve transaction speed, to enhance security, or to transfer control from the owner of the terminal to the owner or holder of the card. In this scenario, what was a series of commands to the card before the installation of the card-side code now becomes one command. The code on the card packages or encapsulates all the previous commands.

When you move code from the reader side to the card side, it is natural to expect that the card services that were available as commands to code on the reader-side are still available as function calls to code on the card side. For example, you will want to be able to read a file on the card-side just as you could read a file on the reader-side. As a result, smart card card-side APIs typically include calls to the functions that implement the command interfaces to cards.

The ISO 7816-4 Standard

Unquestionably the most widely talked-about smart card API is that set forth in the ISO 7816-4 standard, which is discussed at length in Chapter 4, "Smart Card Commands." There are 18 basic interindustry commands on this API:

```
ReadBinary(byte fileId, short offset, byte buffer[])
WriteBinary(byte fileId, short offset, byte buffer[])
UpdateBinary(byte fileId, short offset, byte buffer[])
EraseBinary(byte fileId, byte offset)
ReadRecord(byte record_number, byte mode, byte buffer[])
WriteRecord(byte record_number, byte mode, byte buffer[])
AppendRecord(byte record_number, byte mode, byte buffer[])
UpdateRecord(byte record_number, byte mode, byte buffer[])
GetData(short mode, byte buffer[])
PutData(short mode, byte buffer[])
SelectFile(byte mode, byte info, byte name[])
Verify(byte mode, byte key[])
InternalAuthenticate(byte algorithm, byte mode, byte challenge[])
ExternalAuthenticate(byte algorithm, byte mode, byte response[])
GetChallenge(byte challenge[])
ManageChannel(byte operation, byte channel_number)
GetResponse(byte response[]);
Envelope(byte buffer[])
```

As far as we know, there is no smart card on the market—nor has there ever been—that implements all the ISO 7816-4 commands with all the generality and capability written into the standard. Nevertheless, the ISO 7816-4 commands have served as a reference model for smart card interfacing and have to at least some degree enabled card interoperability. Many off-the-shelf smart cards implement some variant of most of these commands.

The GSM 11.14 Standard

A *subscriber interface module* (SIM) is a smart card that is inserted into a GSM (Groupe Spécial Mobile/Global System for Mobile Communication) cellular telephone. As its name implies, a SIM carries, among other things, the subscriber's account information. The European Telecommunications Standards Institute (ETSI) has published a number of standards covering SIMs and their relationship to the GSM phone.

Unlike the master/slave relationship between terminal and card mandated by ISO 7816-4, GSM 11.14—the ETSI standard describing the interface between the phone and the card—allows for the SIM to initiate communication to the phone. Thus the code running in a SIM card has two APIs: one looking inward to services on the card itself and one looking outward to services on the phone.

The inward-looking API on a SIM card is similar to but not identical to ISO 7816-4. What distinguishes a SIM card from other smart cards is the file system and the encryption algorithms used to authenticate keys.

The outward-looking API on a SIM card contains the following functions:

- `DisplayText`—Displays text on the phone's display window
- `GetInKey`—Gets one key hit from the phone's keypad
- `GetInput`—Gets a string of characters from the phone's keypad
- `MoreTime`—Prevents phone timeout by asking for more processing time
- `PlayTone`—Plays an audio tone in the earpiece and on the phone line
- `PollInterval`—Sets time between `STATUS` commands from phone
- `Reset`—Notifies the phone of changes in the SIM
- `SelectItem`—Sends a selection list to the phone to get user selection
- `SendShortMessage`—Sends a short message to the network
- `SendSS`—Sends a supplementary service request to the phone
- `SendUSSD`—(Not currently defined)
- `SetUpCalls`—Sets up a call on the network
- `SetUpMenu`—Sends a user-selection menu to the phone, which displays it

Notice that the outward-looking API enables the smart card to contact an arbitrary node on the telephone network by using `PlayTone` to communicate with it.

The Vendor APIs

Because card-side programming is still very new, most of the card-side APIs are vendor specific, although most of them include the same fuctionality found on the standards-based APIs. All necessarily support loading of applications onto the card after the card has been manufactured and personalized. MULTOS, DKCCOS, and SPYCOS support secure loading schemes.

Schlumberger's Customer-Oriented System

The operating system in Schlumberger's Multiflex smart card includes the ability to load machine code into and execute machine code from the card's EEPROM memory. This executable code can extend or replace executable code in the Multiflex ROM. Schlumberger calls this functionality the Schlumberger Customer Oriented System (SCOS). Due to security considerations, primarily the fear of the creation of counterfeit cards, the Multiflex card included in this book has had the SCOS capability deactivated.

SCOS is a very powerful capability. It lets you create a custom smart card of your own design without incurring the time, expense, or risk of making a ROM mask, manufacturing the chips, and embedding the chips into cards.

The SCOS toolkit or API consists of 71 general-purpose smart card functions. These functions run the gamut from simple arithmetic and file system access to security functions. Using these functions with your own assembly language code, it is relatively easy to add new commands to the Multiflex card. Nevertheless, keep in mind the preceding discussion about RAM utilization and make sure you understand how each function you call and each function it might call uses RAM.

In order to activate your new command(s), you have to override some part of Multiflex command processing loop. SCOS lets you override any of the following elements of this loop:

- *The card manager.* You would override the card manager if you wanted to take complete control of the card and have it only recognize your new commands. When you override the card manager, you are responsible for all aspects of the card's operation including RAM initialization, transmit function activation, answer-to-reset, and command processing.

- *The receive function.* Code that overrides the receive function would examine each command coming from the terminal and call new command processing code when a new command was detected. If an existing Multiflex command was detected, the receive function would pass it to the Multiflex card manager for processing as usual.

- *The transmit function.* You override the transmit function when you want to alter the response of or data returned from an existing or extended command.

- *The I/O interrupt function.* Your code gets control whenever the microprocessor throws an I/O interrupt. This feature is useful when you want to achieve tight cooperation between the host and the smart card, which is not typically the case.

MULTOS

The MULTOS API is a card-side application programming interface on a smart card running the MULTOS multiapplication operating system. The API provides write-once-run-anywhere functionality to applications written for MULTOS-conformant smart card platforms. MULTOS provides application segregation via the use of ITSEC E6 certified firewalls. MULTOS can support both online applications, where a direct connection between the host application and the card is required, and offline applications, where such a connection is not required. MUL-TOS also provides a secure end-to-end application load/delete facility under card issuer control, even after the cards are in the field.

The description and specifications of the MULTOS API are openly available for license from MAOSCO Ltd. The license gives the developer the right to develop one or many applications and is royalty free. All licensees receive a copy of the MULTOS Application Programmers Reference Manual, which gives details of the API, application programming language (called MEL, the MULTOS Execution Language), and a description of the MULTOS virtual machine. A small license fee (at the time of this writing approximately £100) is payable to cover the cost of production and distribution. If you want a copy, contact Nick Habgood, the CEO of MAOSCO Ltd., at `Customer.Service@MULTOS.com` to become a licensed MUL-TOS application developer.

Unlike the proprietary Java Card API, which is controlled exclusively by Sun Microsystems, the MULTOS specification is controlled, on behalf of the smart card industry and smart card application developers, by a consortium of companies including Motorola, Hitachi, Siemens, Gemplus, MasterCard, Mondex, DNP, and Keycorp.

The Java Card

In October 1996, Schlumberger announced the implementation of an interpreter (virtual machine) for a subset of the Java programming language (Java Card) on a standard 8-bit smart card microcontroller. For the first time in the 20-year history of smart cards, it was possible for anybody to write a card-side application for a smart card and to write it in a modern high-level language.

While the Java Card smart card only supports a subset of the Java language and Java programs running on a smart card that are still time and space constrained, viable smart card card-side applications—particularly non-financial applications—can be quickly and easily programmed and field tested. Providing quick time-to-market for new and innovative smart card applications was one of the driving forces behind the development of the Java Card smart card.

There are no 32- or 64-bit data types, no Unicode characters, no threads, no multidimensional arrays, and no garbage collection in the Java Card smart card. The best way to think of a Java program on a Java Card smart card is as a script around native function calls. It contains processing logic and it passes parameters and collects results, but rarely does a Java program on a Java Card smart card do any real computation. A Java Card smart card program is more of a data flow controller than a data processor.

Java Card API Development

The application programming interface (called the *class library* in Java-speak) provided on the first commercial Java Card smart card, Schlumberger's Cyberflex 1.0, was modest albeit useful and had very much an ISO 7816-4 flavor. This card went on sale in a Prerelease developer's series in May 1997.

Almost immediately, Java aficionados and object-oriented smart card developers started to propose improved class libraries for the Java Card. Java Card class libraries on a smart card are very size constrained. The byte codes of the class library itself displace applications and user data in NVM and the native functions needed to support the library compete for space in ROM. The API design preferred by many smart card vendors and application writers was a small core of services available on every Java Card smart card with industry- and application-specific extensions available on some Java Cards smart card. The Java Card Forum, a consortium of both card manufacturers and card issuers, participates heavily in the effort to define the Java API core services. Consult `http://www.javacardforum.org` for the latest news on this activity.

Sun Microsystems, through its JavaSoft subsidiary, is also very active in class library definition with its Java Card 2.0 API, which it released in October 1997. The JavaSoft Java Card 2.0 API encompasses more of the environment than simply an API, since it defines how an application must be structured, how it must be written, how it must handle communication with the terminal, and what its security policies must be. There is to date no commercially available smart card that implements the entire JavaSoft Java Card 2.0 specification. The details and documentation for the JavaSoft API can be found at `http://java.sun.com/products/javacard`.

Visa has announced that it will base its Open Technology Platform smart card on the Java Card specification, so you can expect that there will be a Visa Java Card class library sometime in the relatively near future. Schlumberger has followed Cyberflex 1.0 with Cyberflex 2.0 Core and Cyberflex SIM, which is a Java Card–based GSM SIM card with its own class libraries. In fact, there is no reason to doubt that card-side APIs will not be buffeted by exactly the same marketplace and technology evolution pressures as reader-side APIs.

One of the nice things about a smart card that contains a virtual machine is that you can load your own libraries onto the card and hence create your own card-side API.

Schlumberger's Java Card 1.0 API

The native function suite underlying Schlumberger's Cyberflex 1.0 Java Card API is a direct descendent of Schlumberger's Multiflex smart card discussed at length in Chapter 6, "Smart Card Software Development Tools." The FlexCash card discussed in Chapter 11, "The FlexCash Card: An E-commerce Smart Card Application," is built using this API and is the ISO-7816 card emulator shown in Listing 8.1.

Listing 8.1. Schlumberger's Java Card version 1.0 API.

```
public class AdminApp
{
  // Constants used throughout the program
  static final byte BUFFER_LENGTH  = (byte)0x50;

  static final byte ACK_SIZE                  = (byte)1;
  static final byte ACK_CODE                  = (byte)0;
  static final byte OS_HEADER_SIZE            = (byte)0x10;
  static final byte GPOS_CREATE_FILE          = (byte)0xE0;

  static final byte ST_INVALID_CLASS          = (byte)0xC0;
  static final byte ST_INVALID_PARAMETER      = (byte)0xA0;
  static final byte ST_INS_NOT_SUPPORTED      = (byte)0xB0;
  static final byte ST_SUCCESS                = (byte)0x00;
  static final byte ST_NO_RETURN              = (byte)0xFF;

  static final byte ISO_COMMAND_LENGTH        = (byte)5;

  static final byte ISO_READ_BINARY           = (byte)0xB0;
  static final byte ISO_UPDATE_BINARY         = (byte)0xD6;

  static final byte ISO_READ_RECORD           = (byte)0xB2;
  static final byte ISO_UPDATE_RECORD         = (byte)0xDC;

  static final byte ISO_INIT_APPLICATION      = (byte)0xF2;
  static final byte ISO_VERIFY_KEY            = (byte)0x2A;
  static final byte ISO_SELECT_FILE           = (byte)0xA4;

  static final byte ISO_DIRECTORY             = (byte)0xA8;
  static final byte ISO_GET_RESPONSE          = (byte)0xC0;
  static final byte ISO_DELETE_FILE           = (byte)0xE4;
  static final byte ISO_GET_FILE_INFO         = (byte)0xF8;

  static final byte GPOS_SET_ACL              = (byte)0xFC;

  static final byte ISO_VERIFY_CHV            = (byte) 0x20;
  static final byte ISO_CHANGE_CHV            = (byte) 0x24;
  static final byte ISO_UNBLOCK_CHV           = (byte) 0x2C;

  static final byte ISO_CLASS                 = (byte)0xC0;
  static final byte ISO_APP_CLASS             = (byte)0xF0;

  public static void main () {
```

```
byte pbuffer[] = new byte[ISO_COMMAND_LENGTH];
byte dbuffer[] = new byte[BUFFER_LENGTH];
byte ackByte[] = new byte[ACK_SIZE];
byte chvbuffer[]= new byte[8];

// I would like to put that on the stack
byte dirbuffer[] = new byte[16];

short fileId;
short offset;
byte bReturnStatus;
byte bTemp;

_OS.Execute((short)0,(byte)0);

/* Initialize Communications */
dbuffer[0]=(byte)0x3B; dbuffer[1]=(byte)0x32; dbuffer[2]=(byte)0x15;
dbuffer[3]=(byte)0x00; dbuffer[4]=(byte)0x49; dbuffer[5]=(byte)0x12;

_OS.SendMessage(dbuffer,(byte)6);

bTemp = 0;

do {
    /* Retrieve the command header */
    _OS.GetMessage(pbuffer,ISO_COMMAND_LENGTH,ACK_CODE);

    // Init vars for iteration
    //fileId = 0;
    //offset = 0;

    /* Verify class of the message - Only ISO + Application */
    if ((pbuffer[0] != ISO_APP_CLASS)
     && (pbuffer[0] != ISO_CLASS)) {
        _OS.SendStatus(ST_INVALID_CLASS);
    }
    else {
      /* go through the switch */
      /* Send the acknowledge code */

      // Verify if data length too large
      if (pbuffer[4] > BUFFER_LENGTH) {
        bReturnStatus = ST_INVALID_PARAMETER;
      }
      else
      {
        switch (pbuffer[1]) {

        //  DeleteFile
        //  This command is not present in the regular bootstrap
        case ISO_DELETE_FILE:
```

Listing 8.1. continued

```
// SelectFile
case ISO_SELECT_FILE:
    /* we always assume that length is 2 */
    if (pbuffer[4] != 2) {
        bReturnStatus = ST_INVALID_PARAMETER;
    }
    else
    {
        //_OS.SendMessage(ackByte,ACK_SIZE);
        // get the fileId in the response buffer
        _OS.GetMessage(dbuffer,(byte)2,pbuffer[1]);
        // cast dbuffer into a short
        fileId = (short) ((dbuffer[0] << 8) |
                             (dbuffer[1] & 0x00FF));
        if (pbuffer[1] == ISO_DELETE_FILE) {
            bReturnStatus = _OS.DeleteFile(fileId);
        }
        else {
            bReturnStatus = _OS.SelectFile(fileId);
        }
    }
    break;

case ISO_GET_FILE_INFO:
    if (pbuffer[4] != (byte)OS_HEADER_SIZE) {
        bReturnStatus = ST_INVALID_PARAMETER;
    }
    else
    {
        // Extract the fileId from P1-P2
        fileId = (short) ((pbuffer[2] << 8) |
                             (dbuffer[3] & 0x00FF));
        if ((bReturnStatus =
            _OS.GetFileInfo(dbuffer)) == ST_SUCCESS) {
            _OS.SendMessage(ackByte,ACK_SIZE);
            _OS.SendMessage(dbuffer,pbuffer[4]);
        }
    }
                    break;

case GPOS_SET_ACL:
    if (pbuffer[4] != 0x08) {
        bReturnStatus = ST_INVALID_PARAMETER;
    }
    else {
        _OS.GetMessage( dbuffer,
                        (byte)0x08,
                        pbuffer[1]);
        bReturnStatus = _OS.SetFileACL(dbuffer);
    }
    break;
```

```
case ISO_DIRECTORY:
    if (pbuffer[4] &gt; (OS_HEADER_SIZE+1)) {
        bReturnStatus = ST_INVALID_PARAMETER;
    }
    else
    {
        _OS.BackupFileStatus();
        _OS.SelectCD();
        // If info asked on current directory, no need to
        // select a record
        if (pbuffer[3] != 0) {
            if ((bReturnStatus =
                _OS.SelectRecord((byte)(pbuffer[3] !=
                ST_SUCCESS){
                    break;
            }
            _OS.InitFileStatus();
        }

        // Get the header information for the file
        bReturnStatus = _OS.GetFileInfo(dbuffer);

        // Reformat the message
        _OS.RestoreFileStatus();
        if (bReturnStatus == ST_SUCCESS) {
            ackByte[0] = pbuffer[1];
            _OS.SendMessage(ackByte,(byte)1);
            _OS.SendMessage(dbuffer,pbuffer[4]);
        }
    }
    break;

// CHV and key manipulation routines
case ISO_VERIFY_CHV:
case ISO_CHANGE_CHV:
case ISO_UNBLOCK_CHV:
case ISO_VERIFY_KEY:
    // Get key number
    _OS.GetMessage(dbuffer,pbuffer[4],pbuffer[1]);

    if (pbuffer[1] == ISO_VERIFY_KEY) {
    bReturnStatus = _OS.VerifyKey(  pbuffer[3],
                                    dbuffer,
                                    pbuffer[4]);
    }
    else {
        // Get a good value for the unblocking flag
        if (pbuffer[1] == ISO_UNBLOCK_CHV) {
            pbuffer[2] = 1;
        }
        else {
            pbuffer[2] = 0;
```

continues

Listing **8.1.** continued

```
            }

            if (pbuffer[1] != ISO_VERIFY_CHV) {
                for (bTemp=0;bTemp<8;bTemp++) {
                    chvbuffer[bTemp] = dbuffer[bTemp+8];
                }
                bReturnStatus = _OS.ModifyCHV ( pbuffer[3],
                                                dbuffer,
                                                chvbuffer,
                                                pbuffer[2]);

                break;
            }

            bReturnStatus = _OS.VerifyCHV(  pbuffer[3],
                                            dbuffer,
                                            (byte)0);

        }
        break;

    case ISO_INIT_APPLICATION:
        /* Should send the id of a valid program file */
        //_OS.SendMessage(ackByte,ACK_SIZE);
        _OS.GetMessage( dbuffer,(byte)1,pbuffer[1]);
        // cast dbuffer into a short
        fileId = (short) ((pbuffer[2] << 8) |
                            (pbuffer[3] & 0x00FF));
        bReturnStatus = _OS.Execute(fileId, dbuffer[0]);
        break;

    case GPOS_CREATE_FILE:
        if (pbuffer[4] != OS_HEADER_SIZE) {
            bReturnStatus = ST_INVALID_PARAMETER;
            break;
        }
        //_OS.SendMessage(ackByte,ACK_SIZE);
        // Receive The data
        _OS.GetMessage( dbuffer,pbuffer[4],pbuffer[1]);
        bReturnStatus = _OS.CreateFile(dbuffer);
        break;

    case ISO_UPDATE_RECORD:
    case ISO_UPDATE_BINARY:
        _OS.GetMessage(dbuffer,pbuffer[4],pbuffer[1]);
        // assumes that a file is already selected
        if (pbuffer[1] == ISO_READ_BINARY) {
            // compute offset from pbuffer[2..3] via casting
            offset = (short) ((pbuffer[2] << 8) |
                                (pbuffer[3] & 0x00FF));
            bReturnStatus = _OS.WriteBinaryFile (offset,
                                                 pbuffer[4],
                                                 dbuffer);
```

```
            }
            else {
                bReturnStatus = _OS.WriteRecord  (dbuffer,
                                                  pbuffer[2],
                                                  pbuffer[3],
                                                  pbuffer[4]);
            }
            break;

        case ISO_READ_RECORD:
        case ISO_READ_BINARY:
            // assumes that a file is already selected
            if (pbuffer[1] == ISO_READ_BINARY) {
                // compute offset from pbuffer[2..3] via casting
                offset = (short) ((pbuffer[2] << 8) |
                                  (pbuffer[3] & 0x00FF));
                bReturnStatus = _OS.ReadBinaryFile (offset,
                                                    pbuffer[4],
                                                    dbuffer);
            }
            else {
                bReturnStatus = _OS.ReadRecord  (dbuffer,
                                                 pbuffer[2],
                                                 pbuffer[3],
                                                 pbuffer[4]);
            }

            // Send the data if successful
            ackByte[0] = pbuffer[1];
            if (bReturnStatus == ST_SUCCESS) {
                _OS.SendMessage(ackByte,ACK_SIZE);
                _OS.SendMessage(dbuffer,pbuffer[4]);
            }
            break;

        default:
            bReturnStatus = ST_INS_NOT_SUPPORTED;
        }
    }

    // Verify we want to return a status
    if (bReturnStatus != ST_NO_RETURN)
        _OS.SendStatus(bReturnStatus);
    }
  }
  while (true);
  }
}
```

This program is available on Schlumberger's Cyberflex Web site (http://www.cyberflex.austin.et.slb.com).

The Java program in Listing 8.1 shows how the Cyberflex 1.0 card can emulate a modest ISO card. It implements the following 13 ISO commands:

```
Select File
Verify Key
Create File
Delete File
Directory
Update Binary
Read Binary
Update Record
Read Record
Verify CHV
Change CHV
Unblock CHV
Verify Key
```

and the following three Cyberflex-specific commands:

```
Execute Application
Get File Information
Set File Access Control
```

The program is just a FOREVER loop that gets an ISO APDU command packet from the terminal, executes it, and returns the status to the terminal.

Schlumberger's Cyberflex 2.0 Core API

In October 1997, coincident with the announcement of Java Card version 2.0, Schlumberger introduced an update of the Cyberflex smart card called Cyberflex 2.0 Core. The idea was to provide a smart card with a Java virtual machine on top of which the Java Card 2.0 API could be built, but one that could also support nonfinancial vertical industry APIs.

Here is a complete list of the 46 functions on the Cyberflex 2.0 Core API:

```
GetMessage
SendMessage
SetSpeed
SendStatus
SelectFile
SelectRoot
SelectCD
SelectParent
CreateFile
DeleteFile
ResetFile
ReadByte
GetFileInfo
ReadBinary
WriteBinary
```

```
SelectRecord
PreviousRecord
NextRecord
ReadRecord
WriteRecord
LastUpdatedRecord
InitFileStatus
BackupFileStatus
RestoreFileStatus
VerifyCHV
ModifyCHV
VerifyKey
GrantSupervisorMode
RevokeSupervisorMode
GetFileStatus
SetFileStatus
SetFileACL
GetFileACL
CheckAccess
GetFileSize
GetRecordLength
GetRecordNb
GetFileType
GetApplicationId
GetIdentity
SetDefaultATR
SendATR
CompareBuffer
AvailableMemory
Execute
ResetCard
```

The following sections describe some of these 46 functions.

Communication

```
GetMessage(byte buffer[], byte expected_length, byte ack_code)
```

This function retrieves a message from the terminal of the expected length and puts it into the buffer. The ack_code parameter is used only for T=0 transmissions. (In T=0 communication, the terminal sends the first 5 bytes of the message and waits for ack_code to know what to do with the remaining bytes of the message. The typical case is when ack_code is equal to 0, in which case all subsequent bytes are transferred and put into the buffer.) By using zero as ack_code, your application works the same with any ISO communication protocol.

```
SendMessage(byte buffer[], byte data_length)
```

This function sends $data\_length$ bytes from buffer to the terminal.

File Management

`SelectFile(short fileId)`

This function selects a file on the smart card and prepares it for access (that is, opens it). This file becomes the *current directory* or *current file*.

`CreateFile(byte file_hdr[])`

This function creates a new file in the current directory with the properties given in *file_hdr*.

`DeleteFile(short fileId)`

This function deletes the named file.

`GetFileInfo(byte file_hdr[])`

This function retrieves information describing the current file.

`ReadBinaryFile(short offset, byte data_length, byte buffer[])`

This function reads *data_length* bytes, starting at byte *offset* from the current file and returns them in *buffer*.

`WriteBinaryFile(short offset, byte data_length, byte buffer[])`

This function writes *data_length* bytes from *buffer* into the current file starting at byte *offset*.

`SelectRecord(byte offset, byte mode)`

This function selects a record number *offset* in a record file. *mode* determines whether the offset to the selected record is taken from the beginning or the end of the file or forward or backward from the current location.

`PreviousRecord()`

This function selects the previous record in the current record file.

`NextRecord()`

This function selects the next record in the current record file.

`ReadRecord(byte buffer[], byte record_number, byte offset, byte length)`

This function returns *length* bytes, starting at *offset* in record number *record_number* of the current record file in buffer.

```
WriteRecord(byte buffer[], byte record_number, byte offset, byte length)
```

This function writes *length* bytes from *buffer* into record *record_number* of the current record file, starting at byte *offset*.

Security

Associated with every file on the Cyberflex card there may be a file of one or more cardholder verification numbers (CHVs) and a file of one or more cryptographic keys. Cardholder numbers are also called personal identification numbers (PINs). PIN files and key files are found either in the same directory in which the file they are protecting is found or in a parent directory of the current directory. The CHV numbers and keys in these files are numbered 0, 1, 2, and so on.

Typically, PIN numbers are four ASCII digits long and let the program perform cardholder operations on the file while keys are 8 bytes long and let the program perform card-owner operations on the file. This is just a convention, however, and you can make PIN numbers and keys be whatever you like and mean whatever you like when you design the security for your card's file system.

```
VerifyCHV(byte CHV_number, byte CHV[], byte unblock_flag)
```

This function compares the 8 bytes stored in CHV with CHV number *CHV_number* in the CHV number file associated with the current file and returns success or failure. If *unblock_flag* is nonzero, the comparison is made to the unblocking CHV number rather than the CHV number indicated by *CHV_number*.

```
ModifyCHV(byte CHV_number, byte old_CHV [], byte new_CHV [], byte unblock_flag)
```

This function changes CHV number *CHV_number* from *old_CHV* to *new_CHV*. If *unblock_flag* is nonzero, the change is to the unblocking CHV.

```
VerifyKey(byte key_number, byte key[], byte key_length)
```

This function compares the *key_length* bytes in *key[]* with the *key_number* key in the key file associated with the current file and returns success or failure. After a number of failures set when the key was created, the key becomes blocked and cannot be unblocked.

```
SetFileACL(byte file_ACL[])
```

This function sets the access control list on the current file.

```
GetFileACL(byte file_ACL[])
```

This function returns the access control list of the current file.

```
CheckAccess(byte operation)
```

This function checks whether the file operation is currently allowed on the current file. Returns success or failure.

Utilities

The following are some examples of the 13 utility functions on the Cyberflex 1.0 API.

```
GetFileSize()
```

This function returns the size, in bytes, of the current file.

```
SetDefaultATR(byte buffer[], byte length)
```

This function sets the default ATR to be the first *length* bytes in *buffer*.

```
SendATR()
```

This function sends the default ATR to the terminal.

```
Execute(short fileId, byte mode)
```

This function terminates execution of the current Java program and begins execution of the Java program in the file *fileId*. The mode byte says whether the card's state (global variables) is initialized.

The JavaSoft Java Card 2.0 API

There are 113 effective functions on the JavaSoft Java Card 2.0 API, organized into 3 families and 17 classes. The classes with their corresponding function counts are

- APPLICATION SERVICES—61 functions

    javacard.framework.APDU—10 functions
    javacard.framework.PIN—5 functions
    javacardx.framework.File—46 functions

- CRYPTOGRAPHY—32 functions

    javacardx.crypto.AsymKey—2 functions
    javacardx.crypto.DES_CBC_Key—6 functions
    javacardx.crypto.DES_Key—2 functions
    javacardx.crypto.RSA_CRT_Key—6 functions
    javacardx.crypto.RSA_Key—5 functions

`javacardx.crypto.Key`—3 functions

`javacardx.crypto.MessageDigest`—1 function

`javacardx.crypto.RandomData`—2 functions

`javacardx.crypto.Sha1MessageDigest`—1 function

`javacardx.crypto.SymKey`—4 functions

- `APPLICATION FRAMEWORK`—20 functions

  `javacard.framework.AID`—3 functions

  `javacard.framework.Applet`—3 functions

  `javacard.framework.System`—12 functions

  `javacard.framework.Util`—2 functions

JavaSoft's Java Card application framework encompasses the loading (onto the card) and execution environment of Java applications destined for a smart card. Each applet is registered with the system when it is loaded onto the card. The framework handles all message traffic between the card and the terminal and calls the application that has been selected by the terminal. The JavaSoft framework is also very similar to the EMV framework, except that the JavaSoft framework handles in card-side code what the EMV framework handles simply by specification and convention.

The Java Card's cryptography interface does not appear to be derived from any known cryptographic API specification, and its compatibility with the Java cryptographic API provided to workstation Java has yet to be defined in detail.

The file interface of Java Card version 2.0 uses 46 functions to deliver much the same functionality as the 11 file interface functions of ISO 7816-4 along with enhanced functionality in other areas.

Summary

When the Intel 8080 was introduced in early 1974, it was designed to be used as a traffic light controller. By the end of 1974, Ed Roberts had noticed that the 8080 had a perfectly general instruction set and introduced the first "home" computer, the MITS Altair. The Altair 8800 had 256 bytes of RAM and an optional 256 bytes of PROM. Later versions ran a virtual machine called Tiny Basic written by a young programmer by the name of Bill Gates.

Smart cards are experiencing a similar transfiguration. Long regarded as simply a secure portable storage medium for a handful of counters—a kind of secure if limited capacity floppy disk—smart cards are slowly coming to be regarded as general-purpose computers and a contender to be the next low-cost computing plateau. Smart card computing is just at its Tiny Basic and Visicalc dawn, however, and the smart card killer app has yet to emerge.

It is interesting to speculate on why it has taken 20 years for the potential of smart card technology to be recognized. The lack of detailed technical information about smart cards has undoubtedly been a factor, as has the difficulty in gaining access to smart card software development tools. This situation was not by chance, but was a result of the deliberate policies and pressures of the big smart card issuers. Controlling information about and access to smart cards was part and parcel of the "smart card security through obscurity" strategy. Any security breach would weaken public confidence in smart cards, so if the goals of the big issuers could be met within a framework of tightly controlled information, then these controls did no harm and possibly precluded an unknown number of security incidents.

This situation was stable as long as smart cards didn't have to make a business case (that is, as long as their use was dictated by a central government in spite of the business case). As soon as smart cards tried to penetrate the U.S. market, where their use was not to be dictated and where they had to compete with much cheaper magnetic stripe cards, it was recognized that only multiapplication cards could make a business case. This meant that as many applications of smart cards as possible had to be brought into being. The only way to do this was to open up smart card application programming to everybody.

Think about a traffic light running Tiny Basic and then think about a Pentium II running Windows NT. There is no reason why this isn't the same distance that we will travel with smart cards—only we'll do it in less than half the time it took microcomputers. Smart card chip manufacturers are planning to produce much more capable smart card chips. Smart card readers are becoming standard equipment on desktop, personal, and network computers. Smart card application builders are starting to get the scent.

At this point, the only thing that is clear is that more and more functionality will be migrating to the card and it will become more and more of a general-purpose computer. Throughout it all, it will hang on to its defining characteristics of being a tamper-resistant, portable personal computer.

CHAPTER 9

SMART CARDS AND SECURITY

One of the primary reasons smart cards exist is security. The card itself provides a computing platform on which information can be stored securely and on which computations can be performed securely. The smart card is also highly portable and convenient to carry. Consequently, the smart card is ideally suited to function as a token through which the security of other systems can be enhanced.

In financial systems, sensitive information such as bank account numbers can be stored on a smart card. In electronic purse applications (cash cards and the like), the balance of some negotiable currency can even be stored on a card. This currency can be credited or debited by external terminals (systems) in a local transaction.

In physical access systems (such as opening the door to your office), a smart card can hold the key through which an electronic system can be enticed to unlock the door and allow entry. In network systems, or even local computer systems, the smart card can hold the password through which a user is identified to the network or local system and privileges are granted by those systems to access information or processing capabilities.

When viewed in the abstract, all these seemingly disjointed systems have very similar needs and operational characteristics, particularly with regard to the security of those systems. This chapter examines some of the general characteristics of systems that are referred to as *security*.

Note

The term *security* is often used in a rather loose fashion to refer to a variety of characteristics related to the performance of transactions between two or more parties in such a manner that everyone involved in the transaction trusts the integrity and, perhaps, the privacy of the transaction. With the advent of computer networks and of highly distributed financial transactions, it is often the case that all the necessary parties to a transaction cannot be physically at the same place, or even at the same time, in order to participate in the transaction.

Consider the purchase of an item with a credit card at an airport gift shop while on a trip. This transaction includes a number of distinct steps:

1. Presentation of the consumer's credit card to the vendor.
2. Validation by the vendor that the cardholder is really the owner of the card.
3. Validation by the vendor that the credit card account represented by the card is valid.
4. Validation by the vendor that the account maintains a sufficient credit balance to cover the cost of the item being purchased.
5. Debiting the credit account represented by the card by the amount of the item purchased.
6. Crediting the account of the vendor with the amount of the item purchased (less any fees due to the bank, and so on related to the credit card transaction).

In the performance of this transaction, the cardholder would also like some assurances that much, if not all, of the information related to the transaction is held private. The credit card name, account number, and validation code should not be obtained by some unscrupulous character bent on making fraudulent purchases with the purloined information.

In the performance of a credit card transaction, there are actually many more components than are mentioned previously. However, in just the steps noted, you can see that physical separation of the various parties to the transaction makes it difficult to guarantee that all these parties are satisfied with the integrity and privacy of the transaction.

This chapter discusses the characteristics of security involved in supporting such a transaction. To facilitate this discussion, the objectives of a security environment are first presented in somewhat abstract terms. Then, some of the elements (we'll call them *players*) of a widely distributed transaction system are examined. Some of the mechanisms currently in wide use to provide the desired characteristics through the identified players are examined. Finally, some of the attacks used to thwart these security mechanisms are reviewed.

Objectives and Characteristics of Security Systems

Security within physical or electronic systems can be viewed as the provision of one or more general characteristics:

- Authentication
- Authorization
- Privacy
- Integrity
- Nonrepudiation

When part or all of these characteristics are provided to the extent required to satisfy all the participants of the transaction, the transaction can be considered secure.

Authentication

Authentication means establishing an identity within a transaction. Consider a very simple (non-electronic) transaction such as a student providing homework to a teacher. In general, the teacher wants to confirm that a specific set of homework comes from a specific student. What's involved in establishing identities in such a transaction? Well, when the homework is turned in to the teacher, the teacher will likely visually recognize the student and accept the homework. In order to identify the homework of a specific student, the teacher may inspect the homework when it is turned in to confirm that the student's name is on it. Later, after the teacher has reviewed the homework and graded the paper, the grade can be recorded next to the name. In such a transaction, an environment of trust must be established. The teacher can associate (visually) a student, the student's homework, and the student's name on the homework, and the teacher believes this association to be true. Establishing this trust environment for a classroom setting is typically subtle and is not usually a rigorous procedure.

The rigor applied to establishing trust is generally commensurate with the value of the transaction. If the transaction does not involve simply homework, but something much more valuable (to one or both parties), such as a final examination or an SAT examination, then establishing the trust environment can be much more involved. Verification of identity may be required at the door of the testing facility; the form of this verification might be a student ID card or a state driver's license. Such forms of authenticated identity suffice to introduce the concept of a trust broker or a trusted third party that both of the parties in the transaction can look to for establishing a trust environment, if they don't know each other. The test monitor may not be able to recognize a student visually, but does know what a valid student ID looks like. So if the student presents such an ID with a picture on it that matches the bearer of the card and a name on it that matches a name on the test list, the monitor can believe that the bearer of the ID card is really the person authorized to take the examination and receive the grade derived from the examination.

If the transaction in question involves something of even greater value (to one or both parties), then establishing the trust environment may be even more involved. For example, purchasing a house with a mortgage loan may require that a wide variety of information be collected and the validity of that information be attested to in legally binding ways.

The object, then, of a security system is to provide authentication mechanisms through which a trust environment can be established among all the participants in a transaction. The participants may not know each other, may not be physically together during the transaction, and may even be participating in the transaction at widely different times (that is, the transaction requires a significant elapsed time to complete).

Authorization

Authorization is the establishment of privileges within a transaction. That is, once the identity of a participant in a transaction has been authenticated, what that participant is allowed to do as part of the transaction must be established. In a financial transaction, this authorization might consist of simply confirming that the authenticated individual has enough money to make the desired purchase or enough money to provide the desired loan. In the earlier exam example, authorization might consist of finding a student's name on the class roster. If the student can authenticate that her identity is Jane Doe and the name of Jane Doe is found by the monitor on the class roster, then that student will be allowed to take the final examination.

Just as in establishing identity (authentication), the length to which various parties in the transaction will go to establish authorization is generally related to value ascribed to the transaction by one or more parties. To gain entry to a room containing particularly sensitive information in a high-security facility, your name might have to be on an access list that can be checked by a guard at that room. To enter the room, you must meet at least two criteria. First, you must present the correct identification information to the guard to establish (authenticate) your identity. Then the guard must find your identity on the list of individuals allowed access to the room.

In some situations, the concepts of authentication and authorization might be merged together. In many office buildings, each office has a physical key. The key patterns may be such that a master key can open any office door. In this case, authentication is established by physical possession of the key. From the standpoint of the lock on the door (which is one of the participants in the transaction of unlocking and opening the door), both the authenticated identity of the individual and that individual's authorization to enter the room guarded by the door is satisfied by that individual physically presenting the key.

Privacy

Privacy is the concept of allowing only the participants in a transaction to know the details of the transaction, and it might even mean that only the participants know that a transaction is occurring.

When a credit card purchase is made, the protocol of presenting the card to the vendor, performing the financial transaction, and returning a receipt of the transaction to the cardholder is set up to minimize the conveyance of sensitive information such as the account name, number, or validation number to those who may be casually observing the transaction. Similarly, when using a telephone calling card at a public telephone, conventional wisdom mandates that one be very cautious to hide the entry of the card number, lest it be seen by someone who will make note of it and use it to make cardholder telephone calls that the cardholder has not authorized.

Integrity

Integrity is the concept that none of the information involved in a transaction is modified in any manner not known or approved by all the participants in the transaction, either while the transaction is in progress or after the fact. In the previous homework example, when the student turns in the homework, the total transaction may not actually be concluded until the teacher reviews the homework and records a grade. In this simple example, the integrity of the information is maintained by the teacher keeping the homework in controlled possession until it is graded and the grade recorded. The student's integrity facility in this case is to get the homework back from the teacher and be able to review it to make sure that it's in the same state as when it was turned in.

For the homework example, the integrity of the transaction system is typically not of paramount importance to the student since teachers don't often maliciously modify homework in their possession. The teacher might be more concerned with the integrity of the information—first, in the sense of knowing that the homework hasn't been modified since it was turned in (usually not too likely), and second, in knowing that the homework was actually done by the student.

This latter aspect is often not guaranteed by any stringent mechanism in the case of homework. In the case of examinations, which might be viewed as more valuable, more proactive mechanisms are sometimes used. For example, some universities make use of an "honor code" under which a student might be required to attest to the fact that an examination was completed by the student and that the student neither gave nor received any assistance during the examination proper. Providing mechanisms to facilitate this concept in the highly dispersed environment of electronic transactions across a wide area computer network is a bit more challenging.

Nonrepudiation

Nonrepudiation is establishing the fact of participation in a particular transaction by all the parties to the transaction, such that none of the parties can claim after the fact that they did not actually take part in the transaction. Mechanisms to facilitate this concept are typically closely related to the mechanisms used to authenticate identity. In many discussions, the two concepts are viewed as essentially equivalent.

Note

Of these five characteristics of security, it is the concept of privacy that precipitates the greatest concerns on the part of governmental entities. As you will see, encrypting information through mechanisms that allow only the intended participants of a transaction to be able to understand it is often a highly regulated capability. The same encryption mechanisms used to establish privacy can often also be used to authenticate identity. When used for authentication, encryption is viewed much more benignly by governmental entities than when used for privacy.

The System Components

The previous section defines some of the abstract characteristics of security as it relates to a variety of transactions. This section defines the components of a networked system; that is, those elements comprising a system through which transactions can be realized. More specifically, this networked system uses smart cards as an integral element of the security infrastructure.

The Card

Smart cards use a computer platform on which information can be stored such that access to it can be strictly controlled by the cardholder, the card issuer, or the provider of any specific applications on the card. Further, software can be executed on the card under strict control of either the cardholder, the card issuer, or the provider of specific applications on the card. Given these characteristics, the smart card provides a variety of useful security characteristics, including

- Storage of passwords for access to computer systems, networks, information stores, and so on
- Storage of keys, public and private, for authenticating identity
- Storage of keys, public and private, for encrypting information to ensure its privacy
- Storage of information to be conveyed to various access points for a system (for example, a financial system) without the cardholder being able to access or change that information in any way
- Performance of encryption algorithms for authenticating identity
- Performance of encryption algorithms for ensuring the privacy of information

The Cardholder

A smart card can represent the cardholder in an electronic environment. Further, the card can be programmed to require some type of identity authentication from the cardholder before it will provide such electronic representation for the cardholder. That is, the smart card can use a variety of mechanisms in a transaction with the cardholder through which the cardholder convinces the card that it should act on the cardholder's behalf. Some of the mechanisms used by the card to authenticate the identity of the bearer include

- Requiring the bearer to enter a personal identification number (PIN)
- Requiring the bearer to enter some known personal information stored on the card
- Requiring some biometrical characteristic of the bearer, such as a fingerprint or a facial image, to be measured by a sensor or collection of sensors and then matched against a benchmark of this characteristic stored on the card
- Requiring the bearer to properly perform a series of operations leading to a specific state known to the card

The identity authentication transaction that occurs between the card and the cardholder is a rather complete specific example of the transaction that one wants to occur generally through the enabling actions of the card. Both sides of the transaction (that is, the card and the cardholder) must be concerned with

- Authenticating the identity of the other (party to the transaction)
- Being authorized with the appropriate privileges once identity is authenticated
- Being assured of the integrity of the transaction
- Being assured of the privacy of the transaction
- Being able to confirm that the transaction took place in a proper fashion

The Card Issuer

The card is typically given to the cardholder by a card issuer. In the case of financial cards, the issuer is generally a bank or other financial institution. The card issuer generally is responsible for providing the system in which the card can function to perform its security-related functions. One aspect of this system is typically the linking of salient information about the cardholder to the functional characteristics of the card. The issuer functions as a certification authority or as a trust broker. It is through the actions of the issuer that various parties of a subsequent transaction can achieve some level of trust in the transaction, although they do not know each other prior to the initiation of the transaction.

In the financial environment, very well-defined protocols have been put in place by associations of financial organizations and buttressed by binding national and international laws and agreements. In the emerging world of computer networks, the existence of equivalent certification authorities is only now being legitimized by evolving system deployment.

The Terminal

The access point of any smart card with any electronic system is typically referred to as a terminal; sometimes the terms *smart card reader* or *smart card interface device* are also used. Terminals can vary significantly in complexity and capability and hence in the level of security that they support. At the most capable level, a terminal is a secure computing platform on par with the smart card itself, although typically not nearly so small, inexpensive, and portable. In such a configuration, a terminal might contain a comparatively powerful computer processor, memory, telecommunications interfaces to local and wide area computer networks, display screens, input devices (for example, a keypad or keyboard) through which a user can enter information (to the terminal's processor and then perhaps on to the smart card), and perhaps even biometric sensors that the terminal can use to ascertain personal characteristics of the cardholder. For example, fingerprint readers and facial characteristics scanners are beginning to emerge within the security marketplace as viable elements of terminals.

A highly integrated configuration including a tamper-resistant computer, memory and secondary storage, and a secure cardholder verification entry facility would typically be provided by a card issuer to a merchant. This terminal provides a secure point of presence (from the standpoint of the card system issuer) in the merchant's environment through which the issuer system can communicate with the cardholder's smart card.

The PC

In emerging computer network environments, the terminal component from earlier smart card–based systems is separated into a computer component (that is, a PC, a network computer, a workstation, or some similar designation) to which is attached a relatively simple smart card reader. This particular configuration raises some security concerns with respect to the use of smart cards. In particular, the cardholder should always understand the security risks in providing verification of identity to the card through a computer system of unknown control. Obviously, this same concern can be raised for all levels of systems; trojan horse ATMs reportedly have been deployed to fraudulently gain account numbers and PINs from unsuspecting users. For home computer–type systems, however, the risks of the system being in a position to capture sensitive information are significantly greater.

The point then is that the cardholder should be reasonably cautious of the computer systems through which the card is used. If it's the personal computer system of the cardholder, then the risks are greatly minimized since the cardholder has control over the system's security environment. If it's a personal computer system being used in a commercial environment, then the cardholder should be concerned with the manner in which a PIN is entered.

For example, if a personal computer configuration makes use of a simple smart card reader and the cardholder is expected to enter a PIN through the computer's keyboard, then it's a relatively simple procedure for the system manager of the personal computer configuration to be able to capture the keystrokes and know the PIN for the cardholder's card. If the computer belongs to the cardholder and is under the direct control of the cardholder, then the security risks (of having the PIN captured) are greatly minimized. For public environments, it is possible to obtain more sophisticated smart card readers which have integrated keypads through which a PIN can be entered and passed on to the card, not to the computer system to which this terminal is connected.

The Network

The network through which computer systems are connected should always be treated as a completely nonsecure environment. The application developer, the card issuer, and the cardholder should all view the communication channel as completely open to the world. Information that passes through these channels can be monitored, captured, and manipulated by unknown persons or systems.

The Application

The application is the particular system or system component that is provided through the auspices of the card issuer (or at least with the concurrence of the card issuer) and is intended to provide some type of service accessed by the cardholder. The application may make use of an infrastructure within the network or within the end computer system through which the cardholder gains access to it. In these cases, however, the application must be concerned with the security of this infrastructure.

In many existing smart card–enabled systems, all the players operate within an environment provided by the card issuer. In the Internet environment, it is more difficult to provide a well-controlled infrastructure for all these players. They must each understand the security limits of the components that they deal with.

The Mechanisms

The previous sections define some of the abstract concepts of security as well as the major components of the systems for which a secure environment is desired. This section examines some of the mechanisms that the various players can use to facilitate the various security concepts.

Physical Security

Central to the overall security architecture is the concept of physical security. The smart card figures very prominently in this. From the cardholder's standpoint, being able to have the smart card computer platform in physical possession is a large step toward overall security. In this case, attacks against the security of the overall system have to be made against the system components while in operation or through examination of information gained while the system was in operation. This means, for example, that attacking encryption algorithms used by the smart card must typically proceed from captured cyphertext, not from active examination of the card while in use.

Conversely, the overall security architecture of the smart card–enabled system must be such that if a card is no longer in the cardholder's possession, the damage to the system through a security attack can be limited through the knowledge that the card is no longer in the cardholder's possession. Further, the vulnerability to the entire system must be minimized if the information related to a single cardholder is compromised.

Processor and Memory Architecture

An adjunct to physical security, at least in the case of the smart card, is the enhanced security architecture of the microprocessor-based computer installed in the card and the tamper-resistant packaging of the card itself. Chapter 2, "Physical Characteristics of Smart Cards," examines the architecture of the smart card's computer. Packaging the processor, memory, and I/O support in a single integrated circuit chip enhances the security of the entire configuration. It is difficult, though certainly not impossible, to connect electrical probes to lines internal to an integrated circuit chip. The equipment to insert such probes is reasonably expensive. Consequently, for an attacker to extract information directly from a chip requires physical possession of the card, costly equipment, and detailed knowledge of both the hardware architecture of the chip and the software loaded onto the chip.

Tamper-Resistant Packaging

The packaging of the integrated circuit chip into a smart card is typically viewed as being tamper-resistant as well as tamper-apparent. *Tamper-resistant* refers to the characteristic that, given physical possession of a smart card, it's a nontrivial task to get to the chip and even more nontrivial to extract information from the chip. *Tamper-apparent*, or *tamper-evident*, refers to the characteristic that to do so will typically leave an obvious trail that the card has been tampered with. Thus, it is difficult to learn the secrets contained within a smart card without the cardholder knowing that the card has been compromised.

Authentication

The field of cryptography is dedicated to the development of mechanisms through which information can be shared in a secure fashion. A variety of mechanisms have thus been developed through which the security concepts discussed earlier can actually be realized. Several different mechanisms have been developed to support the authentication of identity among widely diverse participants in a transaction. A few of the more prevalent of these mechanisms are described in the following sections.

Symmetric Key Authentication

Most, if not all, authentication mechanisms involve the sharing of a secret among all the participants in a transaction. Two such mechanisms involved distinct forms for encryption and decryption of information; the first makes use of a symmetric key encryption algorithm and the second a public-key encryption algorithm. Both of these mechanisms involve a shared secret; however, the manner in which the secret is shared in each case makes the two mechanisms preferable in different situations. Specifically, symmetric key algorithms are most useful in providing bulk encryption of information since they are less processor intensive than public-key algorithms.

Symmetric encryption algorithms make use of a single key for both the encryption and the decryption of information. This is illustrated in Figure 9.1.

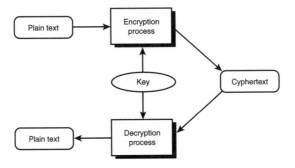

Figure 9.1. *Symmetric key encryption.*

In a symmetric key approach, the same key is fed into the encryption algorithm to both encrypt information and decrypt information. Plain text information is passed into the encryption process, where it is modified through the application of the key value. The resulting cyphertext contains all the information present in the original plain text; however, due to the manipulation of the encryption algorithm, the information is in a form not understandable by a reader that does not possess the key. When the cyphertext is passed back through the encryption algorithm with the same key applied (as was used for the encryption process), the plain text is recovered.

It is apparent that this approach can be used to keep secret the plain text information from anyone who does not have the required key. The approach can also be used, however, to allow each side of a pair-wise transaction to confirm that the other side holds the same key and thereby authenticate a known identity.

This symmetric key identity authentication for a smart card environment is illustrated in Figure 9.2.

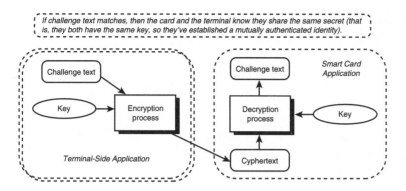

Figure 9.2. *Authentication via shared secret.*

In the case shown in Figure 9.2, the application spans both the terminal-side environment and the card-side environment. In most common instances today, the application is created by the card issuer, who installs the shared secret (the key) in both environments. It should also be pointed out that the case shown in Figure 9.2 could be extended to make use of two distinct authentication operations, each using a different key. This approach would be quite useful if, for example, each of many different cards with different cardholders simply needed to authenticate a single identity for the terminal-side application; in this scenario, the terminal-side application would need to authenticate the unique identity of each individual card.

With this approach, each card would need to know two keys: one to be used to authenticate the terminal-side application and one to use in authenticating itself to the terminal side application. The terminal-side application, however, would need to know a large number of keys: the one it uses to authenticate itself to all the various cards and one for each card it uses when each card authenticates itself to the terminal-side application. From the standpoint of the cardholder, this is a less-than-optimal situation in that the secret key used to authenticate the identity of the card is known outside the card. Similarly, the secret key used to authenticate the identity of the terminal-side application is known to every card that can access this application. If this key can be retrieved from the card, then perhaps an attacker could use it to gain access to the application.

A shared secret approach is typically used by a smart card to authenticate the cardholder to the card. This is done through a cardholder verification (CHV) command set, listed among the inter-industry commands defined in the ISO/IEC 7816-4 specification (see Chapter 4, "Smart Card Commands"). This verification process is often referred to as PIN entry or PIN checking. In this process, a file is written within the file structure of a smart card. A PIN value is then written in this file. When a cardholder inserts a card into a terminal, the terminal-side application requests the bearer to enter a PIN through a terminal keypad. The number sequence then entered is passed through an APDU command to the card which then compares the value supplied by the terminal (which was entered by the cardholder) to the value in the CHV file. If they match, the cardholder is then identified (as far as the card is concerned) as the person for whom the card will act.

This CHV process can be significantly more complex than depicted here; multiple PINs can be required or supported by a single card. In addition, multiple steps can be required for the cardholder finally to be authenticated to the card.

Asymmetric Key Authentication

A second approach that is widely used for identity authentication makes use of an encryption process called *asymmetric key encryption*, commonly called *public-key cryptography*. As the name implies, the technique makes use of different keys for the encryption operation and the decryption operation. This is illustrated in Figure 9.3.

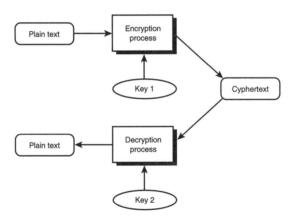

Figure 9.3. *Asymmetric key encryption.*

With this technique, one key is used to encrypt information and the other is used to decrypt the cyphertext to get back the original plain text. There is actually a shared secret between the encryption and decryption keys; that is, the keys are generated as different components of a common algorithm. What's so interesting about this approach, as opposed to the symmetric key approach, is that the proliferation of keys alluded to in the symmetric key discussion can be greatly reduced.

Of the two keys used in the asymmetric key mechanism, one can be held as a closely guarded secret (and indeed should be held as a closely guarded secret) by the entity which will subsequently use this key for identity authentication. This key is termed the *private key*. The other key can now be distributed to anyone; hence, it's referred to as the *public key*. The public/private key pair can be used to establish authenticated identity, as illustrated in Figure 9.4.

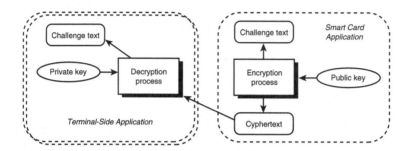

Figure 9.4. *Asymmetric key authentication.*

What's shown in Figure 9.4 is actually just authentication in one direction; that is, the terminal-side application can authenticate its identity to the smart card-side application. To do this, the terminal-side application keeps secret the private key, which essentially represents its identity. The public key corresponding to this private key can be freely distributed; hence, it can be stored on every smart card that might want to access the terminal-side application. So the card can use this public key to encrypt some type of challenge text string; that is, some arbitrary string that the card knows and wants to see if the terminal can find out. It then sends the cyphertext generated from this challenge text to the terminal-side application. If the terminal-side application really possesses the identity (private key) corresponding to the public key possessed by the card, it can decrypt the challenge and return the plain text to the card for validation.

Obviously, the inverse of this procedure can be used to allow the card to authenticate its identity to the terminal-side application. In this case, the terminal-side application needs to know only the public-key component related to each card's identity. If it is to be able to authenticate a large number of cards, it needs to keep a list of, or at least be able to obtain, the public keys of the cards it wants to authenticate. This is considerably better from a security standpoint than the symmetric key situation described in the previous section.

Of course, there's never a free lunch. Computationally, a public-key encryption or decryption algorithm is much more intensive than a symmetric key algorithm. Therefore, it is unrealistic to think about encrypting or decrypting large amounts of information on a smart card processor in a short time. However, the challenge text noted previously does not have to be too voluminous. So for establishing authenticated identity, public-key mechanisms can be used effectively, even given the limited processor capacity of a smart card.

Integrity

The mechanisms described in this chapter are useful for authenticating identities among a variety of parties to a transaction. Although it may not always be thought of as a transaction, the same mechanisms are useful for establishing identities related to documents or procedures; that is, performing the function of allowing an identity to sign a document or a process. This identity authentication procedure can form part of what can be referred to as a *digital signature*. The other aspect of a digital signature is the confirmation that the information which has been digitally signed has not been modified from the time that it is signed until it is read (and the signature's identity is authenticated). This, then, is a means of addressing another of the concepts of security—integrity.

One-Way Hash Codes

In the case illustrated by Figure 9.4, when the smart card encrypts the challenge text with a public key, the smart card authenticates the identity of the terminal side application when it demonstrates possession of the private key which can decrypt the challenge text. The smart card application knows that only the possessor of the private key can decrypt that message. So if the process is reversed and the terminal-side application generates some piece of text and encrypts it with its private key, the smart card application knows that the text, when decrypted with the public key (of the terminal-side application's identity) must have come from the terminal-side application. Thus, the terminal-side application has digitally signed the text in question; that is, it has essentially affixed an identifying symbol that conveys the same information that the signature at the bottom of a contract does.

As mentioned previously, public key encryption and decryption can be very processor intensive. Further, the public-key encryption and decryption operations being discussed are (for purposes of the current discussion) intended to authenticate identity, not assure privacy. This being the case, it is not actually necessary to encrypt all the information in question in order to sign it digitally and to validate that it hasn't changed; that is, that the integrity of the information has been preserved. Rather, all that is necessary is to calculate some type of unique check sum over the information in question and then encrypt that check sum. A *check sum* is the result from a computational algorithm acting on the information in question such that if a single bit of that information changes, the resulting check sum will change.

Generation of such check sums is possible with a family of computational algorithms known as *one-way hash functions*. Through these functions, you can process a large collection of information and derive a much smaller set of information, referred to as a *hash code*. You might think of a hash function as a logical, nondestructive meat grinder. When you grind a piece of meat with it, you don't destroy the meat, but you get pile of ground round that is unique for every piece of meat you put into it. So if I put the same piece of meat through the grinder twice, I get two identical batches of ground round.

Some very desirable traits of useful one-way hash functions are the creation of a unique hash code for a unique collection of bits comprising the source document and the inability (or at least great computational difficulty) to predict what hash code will be generated (without passing the information in question through the one-way hash code algorithm) from a given collection of bits. Two of the more popular one-way hash codes in use in the smart card world are the MD5 algorithm and SHA-1 algorithm.

A variant of one-way hash functions has been developed which also requires a key, in addition to the information in question, before a one-way hash code can be computed. These functions are referred to as *message authentication codes*, or MACs. They are useful for confirming integrity of information as well as authenticating identities associated with the information, but without guaranteeing the privacy of the information. A variant on this theme involves the use of public-key cryptography in conjunction with one-way hash functions. This is the mechanism that is most generally called a *digital signature*, as opposed to the definition suggested earlier.

Digital Signatures

If a one-way hash function is computed on a collection of information and that hash code is not encrypted with the private key of a public/private key pair, the encrypted information that results provides both authentication of the identity of the entity which encrypted the hash code (essentially signed the original information) and the integrity of the original information. This mechanism is illustrated in Figure 9.5.

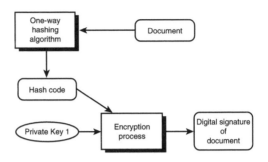

Figure 9.5. *A digital signature using a one-way hash code.*

From Figure 9.5, you can see that when the original document—along with the digital signature of that document—is now passed to another entity, that entity can validate the digital signature. The entity authenticates the identity of the entity that digitally signed the original document and confirms the integrity of the original document (that is, it confirms that the same one-way hash code is calculated from the document on receipt as was calculated from the document at signing time).

Chain of Trust

Through the use of public-key cryptography mechanisms, it is certainly possible to address the concepts of authentication and integrity in a highly dispersed security infrastructure. In fact, many of the same techniques can be used to address authorization and privacy as well, and those points are discussed in the section "Authorization," later in this chapter. At this point, however, an additional issue related to public-key cryptography needs to be addressed. Specifically, how does the recipient of information encrypted with a private key—and which must be decrypted with a public key to authenticate the identity associated with the private key—actually make a connection between the public key and other forms of identification related to that entity?

More to the point, how does the recipient come into possession of the public key and, if a public key would be useful in authenticating the identity of Jane Doe, why would the recipient trust the connection of that public key to the identity of a specific Jane Doe? In more general terms, what is the trust model associated with the use of public-key cryptography?

Certifying Authorities

Most trust models in this emerging infrastructure are based on the concept of a certificate which ties real-world identification information for an entity together with a public-key component of a public/private key pair to be used to authenticate identity in an electronic environment. A certificate is to be issued by a certificate authority, which is some person or entity that will attest to some degree to the connection of identity information to a public key.

One variant of this model makes use of the trust between individuals who know each other to build a chain of trust from one individual to another when the two may not actually know each other. In this model, one receives a public-key and associated identity information from a person they know and who will vouch for the information received. This model could be seen to work for relatively small numbers of individuals, but its applicability for handling very large numbers of individuals is still being explored.

Large-scale trust models are currently rooted in the concept of a certifying authority or even hierarchies of certifying authorities. That is, organizational entities known as certificate authorities (CAs) perform the service of validating identity information, associating that information with a specific entity (person or organization), and associating all this with a public key. This attestation is provided in the form of a document (a certificate) that is digitally signed by the CA. The intent is that the CA forms a trusted third party to all two-way transactions. If two different parties can each trust the CA, then they can trust the information received from the CA (certificates) and hence can trust each other if each has received a certificate from the CA.

Certificates

A certificate is a set of information that connects a physical identity (for example, a name, an address, a telephone number, a Social Security number, a driver's license number) with the logical identity represented by a public/private key pair. Here's an example:

```
Serial Number = 889fba340000000000010000000000
X.509 Certificate Signature Algorithm ID:
    { 1 3 14 3 2 13 } == SHA-WITH-DSA-SIGNATURE
X.509 Certificate Signature Algorithm parameters:
30 5a 02 20 c2 0a 28 7b f5 7e ce 13 c2 a3 6e 72 92 c7 13 67
d9 8f 15 73 e2 ea 19 b1 67 8f 80 f8 8a d4 c2 a3 02 14 ff 9a
ff a2 7b 05 01 2e 99 a8 49 a8 cb 7f d6 ab fd 68 2f 1d 02 20
c0 c9 2d 97 f5 28 11 f5 3b 8d 81 8c 02 59 67 2a 54 25 4b 81
ae 91 c3 70 f9 9b 90 cb de f3 2b 9e
Issuer Name: /C=USA/O=SmartCommerceCorp
Not Valid Before:   12:39:16, 08/30/1997 GMT
Not Valid After:    12:39:16, 11/28/1997 GMT
Subject Name: /C=USA/S=NY/L=Albany/O=SmartCommerceCorp/OU=Sales/CN=Jane
Doe/T=Sales Manager
```

```
Subject Public Key Algorithm:    { 1 2 840 113549 1 3 1 } == Diffie-
Hellman
Subject Public Key Algorithm parameters:
Diffie-Hellman Modulus (p):
575e67ece4e0a0b76fd457621dca50b3fd631c7d622105a3461865da39a42ffb
Diffie-Hellman Generator (g): 6b4b0d3255bfef95601890afd8070994
Subject Public Key: Diffie-Hellman public value =
3bf531a6602de246927003d0121d57d9cf089dbafcc99e65524d40adf73b12aa
```

A variety of recognized standards are associated with such certificates. The information content is defined in the X.509 specification. Actual formats for conveying certificates are defined in the Public Key Cryptography Specifications (PKCS). The specification PKCS #10 defines the format for requesting a certificate from a CA and the specification PKCS #7 defines the format for the certificate issued by the CA.

Authorization

Once the identity of an entity is established through some authentication procedure, what that entity is allowed to do in the context of a given system is the subject of another security concept termed *authorization*. It is useful to think of authorization in the context of a server that is being accessed by a client. The server provides information or some other abstract service to the client, based on what privileges the client has with respect to that server. The model is illustrated in Figure 9.6.

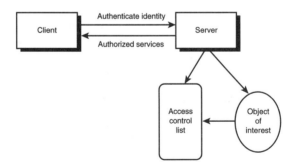

Figure 9.6. *The authorization model.*

This model indicates that if a client wants to gain some type of access to an object of interest, it must first authenticate its identity to the server. The server can then consult a list of privileges that the client (identity) has with respect to that object of interest. The figure denotes this list of privileges as an access control list (ACL); most such mechanisms can be abstracted back to something that looks like an

ACL. This mechanism is widely used within smart cards for access to information (in files or objects) or processing capabilities (functions).

Access Control Lists

Access control lists are authorization or privileges lists which link identities, and what those identities are allowed to do, to specific objects of interest. The ACL is typically viewed from the perspective of the object of interest and the server that makes that object of interest accessible to clients. It is the server that interprets the ACL and decides, based on the authorizations it finds there, what access to allow to the object of interest by the client. For a particular object of interest, a well-defined set of authorizations can typically be defined. For a file, for instance, the following privileges are typically defined:

- Create
- Delete
- Read
- Write
- Modify
- Execute

An ACL for such a file might then look as shown in Table 9.1.

Table 9.1. The access control list for the file `abc.txt`.

Identity	Create	Delete	Read	Write	Execute	Control
Jane Doe	×	×	×	×	×	×
Good Person			×	×	×	
Iffy Person			×			

This ACL says that the identity Jane Doe can do anything at all to the file. The identity Good Person can read, write, and execute any code found in the file while the identity Iffy Person can only read the contents of the file.

Capabilities List

A relatively orthogonal way of looking at this same authorization model (that is, one represented by an access control list) is called a capabilities list. In most instances, the way this variant of the model is implemented, the capabilities list is passed along to the server essentially merged with the identity authentication. That is, there is assumed to be an administration function that decides, external to the actual server, what capabilities (privileges) a specific identity is to have with respect to the object of interest.

In both variants, the security procedures followed are essentially the same; first authenticate the identity, and then go to an authorization list to determine what privileges that identity has with respect to the object of interest.

Privacy

The final concept of security to be dealt with is privacy, which is keeping the details of a transaction secret from everyone not involved in the transaction. The cryptographic mechanisms previously discussed are adequate to provide transaction privacy. In general, the major design factor (that is, deciding which mechanism to actually use) is one of performance in the actual operational environment.

As mentioned previously, public-key cryptography is significantly more processor intensive that is symmetric key cryptography. Consequently, most systems make use of symmetric key algorithms to encrypt the information flow between two disparate points involved in the same transaction. Actually, however, public-key mechanisms are still quite useful in even this case. Specifically, public-key mechanisms are useful in order to exchange the symmetric key needed by both ends of the communication channel. Such shared secrets are well-recognized risk areas in security systems. The longer and more often that the same symmetric key is used, the better chance for an attacker to figure out what it is and use that knowledge to compromise the privacy of the transaction channel.

If public keys are well-known throughout the specific security system, then the mechanisms discussed earlier (in which one end of the transaction channel can generate a random symmetric key and send it, encrypted by the other end's public key, knowing that only the other end possesses the private key necessary to decrypt the message containing the secret symmetric key) can be used.

If public keys are not well-known throughout the system, or even if they are, another mechanism exists which is useful specifically for distributing secret symmetric "session" keys (so named because a new key can be generated for each session in which a transaction of some type is to occur) among disparate participants in the transaction. The mechanism is known as the *Diffie-Hellman protocol*. It has the nice feature that by agreeing in advance to make use of a common algorithm—each end of the channel can calculate a secret key based on information that they can exchange in the clear. The mechanism cannot be used actually to encrypt information, but rather just to exchange a secret symmetric key which can then be used to encrypt the actual transaction information.

Bulk Encryption

The encryption of transaction information is often referred to as *bulk encryption*. In general, smart cards are not involved in bulk encryption processes. The data transfer rate across the I/O port from a smart card reader to the card is very low (on the order of 10 Kbps) relative to typical transmission speeds across local area or even wide area networks. Consequently, most cryptographic operations that are actually performed on a smart card are related to establishment of identity.

The dominant algorithms used for bulk data encryption include the DES algorithm, the Triple-DES algorithm, the RC4 algorithm, and the IDEA algorithm.

Summary

This chapter covers the abstract concepts that are collectively referred to as security. It lists characteristics of security and discusses various mechanisms for realizing those characteristics. The chapter also discusses the role played by smart cards in these various concepts.

PART III

SMART CARD APPLICATION EXAMPLES

CHAPTER

10

THE SMART SHOPPER SMART CARD PROGRAM

In this chapter, you learn about a smart card merchant loyalty program using the 3K Multiflex smart card that comes with this book. This chapter begins by describing the program and the requirements its issuers have set for it. After that, the chapter describes and analyzes the data layout and access control architecture of the program's smart card. Finally, the chapter walks through the specific parts of the application program that deals with the smart card.

The Story of Smart Shopper Cards

A *loyalty program* is a market development technique that generates repeat business by providing customers with rewards based on the total amount of business they do. Frequent buyer programs, such as the American Airlines AAdvantage Miles program, are examples of loyalty programs. Loyalty programs are very effective for merchants and very popular with customers.

As loyalty programs have grown in popularity, they have also grown in complexity and customer expectations. A small- to medium-sized merchant who decides to run a loyalty program may discover that there are many details of running an effective program with which he has neither the time nor the knowledge to deal. Furthermore, he will be interested in sharing the expense of running the program with other merchants who don't compete with him. Thus was born the Smart Shopper loyalty program and the Smart Shopper card.

Smart Commerce Solutions Inc. is the imaginary card issuer for a multiple-merchant smart card loyalty program called Smart Shopper. Here's how Smart Shopper works: Consumers join the program by purchasing a Smart Shopper card from any participating merchant. The Smart Shopper card comes in two sizes—a 5-merchant card that costs $15.00 and a 15-merchant card that costs $30.00. Smart Commerce Solutions is using Schlumberger's 3K Multiflex (just like the one included in this book) for the 5-merchant card and Schlumberger's 8K Multiflex card for the 15-merchant card. If the consumer participates in only a few, selected loyalty programs, she would purchase the 5-merchant card. If the consumer is a professional shopper who collects frequent buyer points wherever she goes, then she would purchase the 15-merchant card.

After purchasing a Smart Shopper card, the consumer can ask the merchant from whom she purchased the card—or any other merchant participating in the Smart Shopper program—to put his loyalty program onto the card. As soon as this is accomplished, the customer is participating in the merchant's loyalty program.

Each merchant loyalty program is allocated 450 bytes of nonvolatile memory on the Smart Shopper card. In this space, the merchant can provide customers with one or more of the prebuilt loyalty schemes provided by Smart Commerce Solutions. The following are examples of the predefined schemes that the merchant can use to define his own loyalty program:

- Frequent buyer points—Accumulate points toward rewards based on the total amount of purchases made with the merchant.
- Cumulative purchasing—Keep track of the total amount of purchases of a particular product in order to qualify for volume purchase discounts.
- Want list—Store lists of desired but infrequently available items such as particular book or video tape on the card that the merchant can instantly match against current inventory.
- Personal preferences—Store personal preferences and requirements such as colors, sizes, and brands on the card so that the merchant can quickly tailor product offerings to the customers' demands.
- Revisit coupons—Store discount coupons on the card that can be redeemed at points in the future to encourage the customer to return.

Besides selecting options such as rewards for total purchases versus frequent visits in the various programs provided by each merchant, consumers purchasing the Smart Shopper smart card can, at their option, enter personal profile data that can be used by all merchants. Such data can, for example, include the cardholder's name, home address, business address, telephone and fax numbers, email address, and even credit card account numbers and sizes.

Merchants Using Smart Shopper

Two merchants participating in the Smart Commerce Solutions Smart Shopper smart card program are Harvest Festival, a grocery store, and Scrivener's Corner, which sells books on the World Wide Web. Both of these merchants have chosen to offer a frequent buyer points loyalty scheme. In addition, Harvest Festival offers the cumulative purchasing scheme and Scrivener's Corner offers the want list scheme.

The frequent buyer points program provided by Smart Commerce Solutions to merchants participating in the Smart Shopper smart card program maintains a running point total of the amount of business the customer does with the merchant. The units of this point total can be dollars, visits, or any other metric chosen by the merchant. The total is incremented after every transaction with the merchant, and the points accumulated can be redeemed with the merchant however the merchant chooses.

The cumulative purchasing program provided by Smart Commerce Solutions and offered by Harvest Festival lets the customer enter a list of individual products for which she wants to track long-term purchasing totals. Each time one of these products is purchased, the cumulative amount of money the customer has spent on that product is increased. The customer may wish to do this for budgeting purposes or to be able to demonstrate her personal buying power in order to negotiate special volume purchases of a product.

The want list program provided by Smart Commerce Solutions and offered by Scrivener's Corner lets customers enter a list of books or book types they are seeking. When Scrivener's accesses the card from its Web server, it will access this list and cross-check it against its current inventory. The Web server will then prepare a page that makes ordering of any hits quick and easy.

Smart Commerce Solutions knows that the success of the Smart Shopper card depends critically on protecting the data security, privacy, and accuracy concerns of all program stakeholders: merchants, customers, and Smart Commerce Solutions itself. As a result, Smart Commerce Solutions has made explicit the security properties of the card program it intends to implement and enforce:

- A lost or stolen Smart Shopper card cannot be used by the person or organization that is not the rightful owner. In particular, no protected data on the card can be viewed or altered.

- It should not be possible to forge a Smart Shopper card. All program software should work only with genuine Smart Shopper smart card program cards.

- It should not be possible to forge Smart Shopper application programs. The Smart Shopper card should not allow itself to be altered or updated by anything other than genuine Smart Shopper application programs.

- The cardholder has complete control over what personal preference information Harvest Festival and Scrivener's Corner can access. The customer should be able to make different personal preference information available to each.

Scenarios for Using Smart Shopper

In order to test the design of the Smart Shopper smart card program, consider the following usage scenarios:

- *Tracking purchases and collecting frequent buyer points*: Alice has just purchased a basket of groceries at Harvest Festival and is at the cash register checking out. She inserts her Smart Shopper card into the smart card reader at the checkout counter. The reader is connected to the cash register and via a network to a server in the back office. When Alice inserts her card, the server reads her cumulative purchasing list off the card. As the items in her grocery basket are scanned, the server cross-checks the universal product codes (UPCs) on Alice's list against the scanned code, and when it finds a match, updates the cumulative purchasing figure on Alice's card with the amount of money spent against the code on this trip to the store. After all the groceries in Alice's basket have been rung up, the cash register updates the frequent buyer total on Alice's card to reflect her purchases during this trip to the store.

- *Using personal preferences and spending frequent buyer points*: Bob inserts his Smart Shopper card into the smart card reader attached to his home computer, connects to the Internet, and picks the Scrivener's Corner home page from his bookmark list. The Java applet sent to Bob's WWW browser off the Scrivener's home page reads the want list off Bob's card and sends it back to the Scrivener's server. The server quickly matches Bob's list against Scrivener's book inventory so that when the home page itself is downloaded to Bob's browser, it includes a handy order form of books on Bob's preference list.

Since all of Bob's billing, shipping, and payment information is also on his Smart Shopper card, all Bob would have to do is check the books he wants and select Done. In this particular case, however, besides the books he wants, Bob also checks the Use Scrivener's Points box, which means that the Scrivener's Corner frequent buyer points total on Bob's card will be set to zero and all the points that were there are applied toward his purchase. After he selects the Done button, the books are on their way.

- *Browsing and editing the Smart Shopper card*: While shopping at Harvest Festival, Alice passes a Smart Shopper customer service kiosk. She inserts her Smart Shopper card and gets a display of each of the merchant programs on her card, together with her current frequent buyer point holdings in each. When she selects the Purchase Tracking button, she gets a display of the products whose cumulative purchasing she is tracking, together with the total amount spent to date on each product. She can use the mouse and the keyboard on the kiosk to delete items and add new items to her purchase tracking list.

High-Level Design

The Smart Shopper smart card program consists of the following major system components:

- *A Smart Shopper smart card.* This card contains Smart Commerce Solutions program administration and security data, cardholder personal data, and merchant-specific programs for each of a number of merchants. The master file of the Smart Shopper smart card contains a file for card administration information, a file for personal data of the cardholder, and a subdirectory of a fixed size for each merchant program. Each merchant directory contains the files needed to implement the schemes selected by the merchant.

- *A program library.* This library includes a number of prebuilt and canned loyalty schemes that the merchants participating in the Smart Shopper program can use to define their individual loyalty programs. In order to encourage the creation of merchant-specific extensions to the Smart Shopper smart card program, Smart Commerce Solutions provides to every participating merchant a high-level application program interface and program library to facilitate the building of Smart Shopper card application programs. This program library can be used to build applications that run on standalone kiosks, that are networked with existing business systems, or that work in tandem with World Wide Web servers.

- *The Smart Shopper Web site.* This site, run by Smart Commerce Solutions Inc., lets customers examine their smart cards from their homes or offices and edit personal data fields. The Smart Shopper Web site provides information and interactive applications for both customers and merchants. Customers can review the contents of their card and update personal information files. Merchants can use interactive utilities at the site to allow customers to update merchant-specific data on the card. Web merchants can send their customers to the site to download their loyalty program into customers' cards.

File Layout

The overall data architecture of the five-merchant Smart Shopper card is shown in Figure 10.1. The master file contains Smart Commerce Solutions administrative data, cardholder personal data, and a subdirectory for each of the five merchant programs on the card.

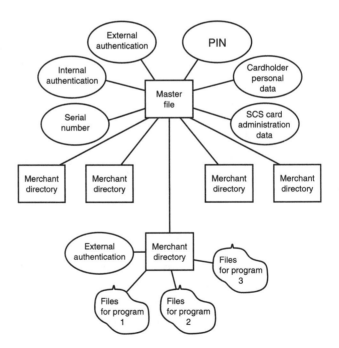

Figure 10.1. *File layout of the Smart Shopper card.*

The Smart Commerce Solutions administrative data file is used primarily for security purposes. Among other things, it contains a directory of the currently active merchant programs on the card.

The cardholder personal data file is under the control of the cardholder and could, for example, contain names, telephone numbers, billing and shipping addresses, and payment information, such as credit account numbers.

Each merchant directory contains an external authentication file with the keys that the merchant uses to administer the security of his own loyalty program. Each merchant directory also contains one or more loyalty schemes which define the merchant's loyalty program. These schemes are provided in a general form by Smart Commerce Solutions Smart Shopper card and are particularized by Smart Commerce Solutions to the needs and requirements of each merchant.

The Master File

The master file ($3F00_{16}$) on the five-merchant Smart Shopper card contains the following files:

- Serial number file (0002_{16})
- PIN file (0000_{16})
- External authentication file (0011_{16})
- Internal authentication file (0001_{16})
- Cardholder personal data (0200_{16})
- Smart Commerce Solutions card administration file (0100_{16})
- Merchant #1 directory (1000_{16})
- Merchant #2 directory (2000_{16})
- Merchant #3 directory (3000_{16})
- Merchant #4 directory (4000_{16})
- Merchant #5 directory (5000_{16})

The serial number, PIN, and external and internal authentication files are discussed in Chapter 5, "The Schlumberger Multiflex Smart Card." The personal identification number in the PIN file is what the cardholder enters to identify herself to the card; it is what the card uses to ensure that it is being held by the authentic cardholder. The external and internal authentication files contain keys that are used by Smart Commerce Solutions to provide security for its administrative functions and data.

A merchant directory (x000$_{16}$) always contains the merchant external authentication file (0011$_{16}$), which contains the keys that the merchant uses to administer security on this loyalty program. A merchant directory also contains one or more of the loyalty schemes provided by Smart Commerce Solutions.

The Cardholder Personal Data File

The cardholder personal data file is a fixed-length record file consisting of 20 records of 24 bytes each. Each record starts with four ASCII characters, which name the cardholder data value contained in the record; that is, NAME for name, SHAD for shipping street address, and so on. These four characters are followed immediately in the record by up to 20 ASCII characters which comprise the data value itself. Here's an example:

```
NAMESally Green
HTEL1 617 484 0391
OTEL1 617 484 3307
SHADOne Main Street
SHCTCambridge
SHSTMassachusetts
SHZP02142
```

The names of all the possible values in the cardholder personal data file, NAME, HTEL, and so on, are standardized by Smart Commerce Solutions so that all the merchants participating in the program can use them. A cardholder is not obliged to enter any values in the personal data file and may enter only those values she is comfortable with providing to the program merchants.

The Smart Commerce Solutions Card Administration File

The Smart Commerce Solutions card administration file contains, among other things, a registry of all active merchant programs currently loaded on the card. This is one way Smart Commerce Solutions ensures that the card is used only by merchants who have joined and are currently active in the Smart Shopper smart card program.

The Merchant External Authentication File

The merchant external authentication file is a transparent file containing three keys of 8 bytes each. These keys can be used by the merchant to define a security architecture and policy for the files in his directory.

The Frequent Buyer Points File

The merchant directory of each merchant choosing the frequent buyer points scheme includes a file that contains the consumer's frequent buyer points total. This is a cyclic record file consisting of ten 6-byte electronic purse transaction records. Each record contains a 6-byte number, and the current record contains the current frequent buyer points total. The result of each credit or debit to the cyclic file is written to the oldest record in the file, and thus the file contains a log of the last ten transactions and balances. If any difficulty is encountered during an update, the file is reverted to the previous valid balance value.

An update transaction can produce erroneous data if, for example, power is removed from the smart card or the card is reset while the new data is being written to the EEPROM of the card. By using a cyclic file, all that is lost is the latest transaction, not the entire point total. It is obviously easier for the merchant and the customer to reconstruct just the latest transaction than to reconstruct the entire purchasing history of the customer.

The Cumulative Purchases File

Besides the frequent buyer points scheme, Harvest Festival has included the cumulative purchases scheme in its loyalty program. The file supporting cumulative purchases is a fixed-length record file that contains a product descriptor followed by a numeric total. The size of the file, length of each record, coding of product descriptors, and units of the numeric total are all defined by Harvest Festival in the process of particularizing the scheme to their requirements. For Harvest Festival, the file consists of 20 18-byte records. Each record contains a 12-digit UPC as ASCII digits and a 6-byte integer that represents the total number of cents the customer has spent on that UPC. Each time the customer purchases a product whose UPC is in the cumulative purchases file, the amount of money spent on the product is added to the integer associated with the UPC. As a result, the consumer can keep track of the total amount of money she is spending on these particular products. She can use this information for budgeting purposes or to bargain for special volume purchase discounts from Harvest Festival.

Volume purchase discounts differ from bulk purchase discounts in that they are given against a promise to buy a specified volume of product over a time horizon rather than all at once. For example, I may be willing to agree to buy two cases of a particular brand of soda pop per week for the next year, but I certainly don't want to take 104 cases home with me today. Volume purchase agreements are stock components of business-to-business commerce. The Smart Shopper card brings this mutually beneficial business arrangement to consumer and retail commerce.

The Want List File

Besides the frequent buyer points scheme, Scrivener's Corner has chosen to include the want list scheme in its loyalty program. The file supporting the want list scheme is a fixed-length record file, each record containing a two-character type flag followed by a descriptive string, followed by a dash and finally a comma-separated list of modifiers. The tags and the vocabulary of the modifiers are defined by Scrivener's Corner in the process of particularizing the general want list scheme. The want list file consists of ten records of 40 bytes each. Each record represents an item in the cardholder's want list. The first two characters tell what kind of wanted item is being described by the record, and the last 38 bytes contain the description itself. Here's an example:

```
AUJane Austen - First Edition, Very Good Condition
TIFor Whom the Bell Tolls - First Edition, Signed
SUGothic Romance - Paperback, Used
```

This is a Scrivener's Corner want list for a customer looking for

- First editions of any work by the author Jane Austen
- Signed first editions of the work titled *For Whom the Bell Tolls*
- Paperbacks in the subject category of gothic romances

The item names such as AU, TI, and SU, together with the qualifier keywords such as First, Folio, and Paperback, have been defined by Scrivener's Corner. Since the cardholder must use a Scrivener's Corner Web page to define her want list and have it written onto her card, the application program behind this page can control the vocabulary used to define want list entries. This ensures each entry conforms to Scrivener's rules and can be understood by the Scrivener's inventory lookup program.

File Sizes

Table 10.1 lists the sizes of the various directories and files on a five-merchant Smart Shopper card with the Harvest Festival and Scrivener's Corner merchant programs loaded onto it.

Table 10.1. File sizes on the Smart Shopper card.

Directory Name	FileId	File Type	Record Size	Records	Data Size	Overhead	Total Size
Master File							
Master File	3F00₁₆	Directory			2910	24	2934
Merchant #1	1000₁₆	Directory			450	24	474

Directory Name	FileId	File Type	Record Size	Records	Data Size	Overhead	Total Size
Master File							
Merchant #2	2000_{16}	Directory			450	24	474
Merchant #3	3000_{16}	Directory			450	24	474
Merchant #4	4000_{16}	Directory			450	24	474
Merchant #5	5000_{16}	Directory			450	24	474
							2370
PIN	0000_{16}	Transparent			23	16	39
Internal Authorization	0001_{16}	Transparent			14	16	30
Serial Number	0002_{16}	Transparent			8	16	24
External Authorization	0011_{16}	Transparent			38	16	54
Personal Data	0100_{16}	Fixed Record	24	10	240	56	296
Administration	$0F00_{16}$	Transparent			105	16	121
							564
Harvest Festival							
External Authorization	0011_{16}	Transparent			38	16	54
Points	1002_{16}	Cyclic	10	6	60	40	100
Cumulative	1003_{16}	Fixed Record	18	12	216	64	280
							434
Scrivener's Corner							
External Authorization	0011_{16}	Transparent			38	16	54
Points	2002_{16}	Cyclic	10	6	60	40	100
Want List	2003_{16}	Fixed Record	24	10	240	56	296
							450

Card Security Architecture

The operating characteristics of the Smart Shopper card—who can see what, who can change what, and so on—are determined by the access conditions placed on the files on the card, together with who possesses what keys stored on the card.

This collection of access conditions together with the key distribution policy taken as a whole is the Smart Shopper card security architecture.

Smart Commerce Solutions has control of the overall security of the card together with control of the files in the master file and is the only entity that knows the keys in the external authorization file in the master file. Furthermore, Smart Commerce Solutions knows one key in the external authorization file in each merchant directory. This key allows Smart Commerce Solutions to block access to the external authorization file itself, if necessary. Blocking this file would essentially deactivate the merchant's program on the card.

By virtue of knowing one key in the external authorization file in the master file, each merchant can create new files in the directory into which his program is loaded and, in the process, can set all the access conditions on these new files. The merchant knows all but cannot change any but one of the keys in the external authorization file in his directory. He can use the keys as he wishes in crafting a security architecture for his program. The merchant cannot delete files in his directory because this would let him delete the external authorization file and remove Smart Commerce Solutions's control over activation of the merchant's program.

Merchants can, however, activate and deactivate the frequent buyer points file. The frequent buyer point total on a Smart Shopper card is a liability for the merchant because he is obliged to exchange these points for other value. As a result, the merchant must have total control over the growth of this liability. If the merchant suspects that a particular customer is receiving unauthorized increases in their frequent buyer total—for example, by working in collusion with a store employee—the merchant can deactivate the frequent buyer point total file to immediately stop the incursion of further liability while the situation is being investigated.

The cardholder has complete control over her personal data. This control is exercised by putting a PIN access condition on all operations that view or change this data. This does not mean that the cardholder must enter or edit by hand all this information. It means that such access is impossible without the cardholder being aware that the access has been granted through the entry of a PIN. Merchant applications have to access this information, and merchants may provide convenient utilities for updating this information.

Table 10.2 lists the details about which entity can perform which actions on which files on a Smart Shopper card with Harvest Festival and Scrivener's Corner programs loaded on it.

Table 10.2. File access conditions on the Smart Shopper card.

Directory Name	FileId	DIRECTORY	DELETE FILE	CREATE FILE	REHABILITATE	INVALIDATE
Master File	$3F00_{16}$	Never	Smart Commerce	Smart Commerce	Smart Commerce	Smart Commerce
Merchant #1	1000_{16}	Never	Smart Commerce	Harvest Festival	Smart Commerce	Smart Commerce
Merchant #2	2000_{16}	Never	Smart Commerce	Scrivener's Corner	Smart Commerce	Smart Commerce
Merchant #3	3000_{16}	Never	Smart Commerce	Smart Commerce	Smart Commerce	Smart Commerce
Merchant #4	4000_{16}	Never	Smart Commerce	Smart Commerce	Smart Commerce	Smart Commerce
Merchant #5	5000_{16}	Never	Smart Commerce	Smart Commerce	Smart Commerce	Smart Commerce

Filename	FileId	READ/ SEEK	UPDATE/ DECREASE	INCREASE	CREATE RECORD	REHABILITATE	INVALIDATE
Master File							
PIN	0000_{16}	Never	Smart Commerce	Never	Never	Smart Commerce	Smart Commerce
Internal Authentication	0001_{16}	Never	Never	Never	Never	Smart Commerce	Smart Commerce
Serial Number	0002_{16}	Always	Smart Commerce	Never	Never	Never	Never
External Authentication	0011_{16}	Never	Smart Commerce	Never	Never	Smart Commerce	Smart Commerce
Personal Data	0100_{16}	Cardholder	Cardholder	Cardholder	Never	Smart Commerce	Smart Commerce
Administration	$0F00_{16}$	Smart Commerce	Never	Never	Never	Smart Commerce	Smart Commerce
Harvest Festival							
External Authentication	0011_{16}	Never	Smart Commerce	Never	Never	Smart Commerce	Smart Commerce
Points	1002_{16}	Cardholder	Harvest Festival	Harvest Festival	Never	Harvest Festival	Harvest Festival
Cumulative Purchases	1003_{16}	Cardholder	Harvest Festival	Harvest Festival	Never	Harvest Festival	Harvest Festival

continues

Table 10.2. continued

Filename	FileId	READ/ SEEK	UPDATE/ DECREASE	INCREASE	CREATE RECORD	REHABILITATE	INVALIDATE
Scrivener's Corner							
External Authentication	0011_{16}	Never	Smart Commerce	Never	Never	Smart Commerce	Smart Commerce
Points	2002_{16}	Cardholder	Scrivener's Corner	Scrivener's Corner	Never	Scrivener's Corner	Scrivener's Corner
Book Want List	2003_{16}	Cardholder	Cardholder	Never	Never	Scrivener's Corner	Scrivener's Corner

Personalizing the Smart Shopper Card

Consumers can purchase a Smart Shopper card at any participating merchant, or on the Smart Commerce Solutions Web site. When the card is purchased at a participating merchant, the merchant can load on the card his own program, but not programs of other merchants. When the card is purchased through the Smart Commerce Solutions Web site, the consumer can load the loyalty programs of any of the participating merchants from the Web site.

Consumers can also, at their discretion, add their own personal data both to the master personal data file in the master file and to any personal data files in the merchant directories. Some merchants may offer this capability as part of the card purchase, but others may not. For example, Harvest Festival doesn't want customers personalizing their new Smart Shopper card in the checkout lane because it would hold up other customers waiting to check out. Harvest Festival does offer a Smart Shopper card kiosk near the customer service window. For those merchants who don't offer this capability and for consumers who'd like to personalize their cards in the privacy of their home or office, Smart Commerce Solutions provides Smart Shopper card personalization forms on the Smart Commerce Solutions World Wide Web site.

The Smart Commerce Solutions Web Site and Smart Commerce Solutions Application Programs

Beside enabling the consumer to purchase and personalize a new Smart Shopper card, the Smart Commerce Solutions Web site contains a number of utilities to browse and update an in-use Smart Shopper card over the Internet.

To use the Smart Commerce Solutions Web utilities, the customer inserts his Smart Shopper card into the smart card reader on his personal computer or into the smart card slot on his network computer. When Smart Commerce Solutions's Smart Shopper home page is contacted, a Java applet is downloaded to the customer's Web broswer. This applet is the connection between the customer's Smart Shopper card and the Smart Commerce Solutions Web server. Figure 10.2 shows the general flow of information between the Smart Commerce Solutions server and the customer's card.

A session begins with the Smart Commerce Solutions applet reading the current data from the Smart Shopper card and sending it encrypted to the Smart Commerce Solutions WWW server. The server then prepares a hierarchy of forms containing this data, which is sent to the customer on demand for editing and updating. When the customer has completed all editing, new and changed data are downloaded back to the Smart Shopper smart card via the Smart Commerce Solutions browser applet.

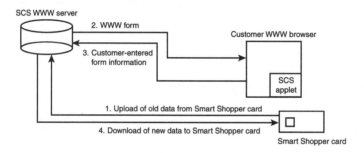

Figure 10.2. *The process of updating a Smart Shopper card on the Web.*

Smart Commerce Solutions maintains the Smart Shopper server as a service to both customers and merchants. Customers can use the server from home or office. Web merchants can point to the server from their own servers. Retail merchants can place Smart Shopper customer service kiosks in their retail locations.

The Smart Commerce Solutions applet is not specific to the Smart Commerce Solutions server, but rather is a general-purpose connection between any WWW participating merchant server and the Smart Shopper card. It accepts commands from an application on the WWW server to read and return data from the card, or to add or update data on the card. Smart Commerce Solutions provides this general-purpose applet to help merchants develop Web-based services and Web-based loyalty schemes.

Smart Commerce Solutions also uses the WWW server to collect data on customer merchant loyalty program preferences. Statistics are gathered both on which merchant programs customers are loading on their cards and which loyalty schemes are most attractive to customers. This information is sent to participating merchants as part of their participation package. All data is summarized just like census data so individual customers cannot be identified. Furthermore, data is sent to the merchant encrypted with the merchant's public key so that only the merchant can decrypt and read it.

Finally, Smart Commerce Solutions distributes on disk and as a download from its Web site a free Windows program that Smart Shopper cardholders can use to browse and edit their Smart Shopper smart cards.

The Smart Shopper Card Browser Program

The Smart Shopper card Browser is a standalone Windows program that lets cardholders view, edit, and update information on their Smart Shopper cards. The program is written in Visual C++ and is provided free to Smart Shopper cardholders by Smart Commerce Solutions. It is built on top of a Multiflex 3K application programming interface library written to PC/SC specifications. The source code for the library and the card browser program are included on the CD-ROM that accompanies this book.

The Harvest Festival Application Programs

There are two Harvest Festival Smart Shopper application programs. The first and primary Harvest Festival Smart Shopper program updates both the frequent buyer points file and the cumulative purchases file (see Figure 10.3). This program is integrated with the check-out line system so that updates to a customer's Smart Shopper card occur as an integral part of normal check-out. The primary application will also allow a customer to use frequent buyer points in paying for a purchase. The second application program makes volume discount offers to the customer based on the cumulative purchasing totals stored on the card. The cardholder can access this program at a kiosk near the customer service window and on the Harvest Festival Web site.

The Scrivener's Corner Application Program

Scrivener's Corner provides pointers from its Web site to the Smart Commerce Solutions site to enable its customers to edit and update the information in the Scrivener's Corner directory. The only application that Scrivener's maintains is one that reads the want list from the customer's card and prepares and transmits the form that matches the customer's want list against Scrivener's inventory.

The Scrivener's want list application uses the general-purpose Smart Shopper Java applet provided by Smart Commerce Solutions to communicate with the customer's Smart Shopper card. First, it sends a command to read and return the customer's book want list in the Scrivener's directory on the Smart Shopper card. Upon receiving the list, the Scrivener's application matches items on the list against current inventory and prepares a standard HTML form page of hits that it returns to the customer browser. This page allows the customer to check off the books he wants to purchase and offers the option of buying with the frequent buyer points stored on the card.

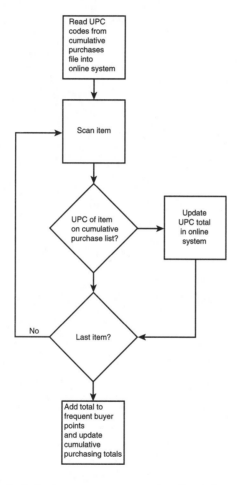

Figure 10.3. *Harvest Festival checkout counter processing for updating the Smart Shopper card.*

If the customer does not elect to use frequent buyer points for this purchase, a commercial "shopping cart" Web purchasing application is activated to complete the transaction. At the end of the transaction, the Smart Commerce Solutions applet is commanded to increase the frequent buyer points total in the Scrivener's directory by the amount of the purchase. If the customer elects to pay with existing frequent buying points, the Smart Commerce Solutions applet is instructed to decrease the frequent buying point total on the card by the amount of the purchase. Figure 10.4 is a flowchart for the Scrivener's Corner Smart Shopper application.

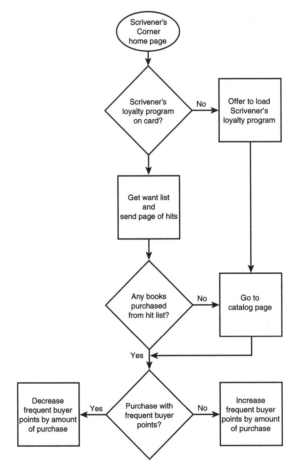

Figure 10.4. *The process of making a Smart Shopper purchase on the Web.*

The Smart Commerce Solutions Smart Shopper Card Management Utility

The Smart Commerce Solutions Smart Shopper card Management Utility is a Windows program that Smart Commerce Solutions gives away on disks and makes available for free downloading from its Web site. The program lets customers update the personal profile data on their Smart Shopper cards and review the frequent buyer points they've received from merchants who have elected to include frequent buyer points schemes in their loyalty programs. The human interface to the Card Management Utility is shown in Figure 10.5.

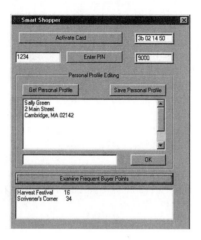

Figure 10.5. *The human interface for Smart Commerce Solutions's Smart Shopper card Management Utility.*

The Smart Shopper card Management Utility is an MFC Visual C++ program that was written against the Multiflex application programming interface described in Chapter 5.

The connection to the Smart Shopper card is established in the OnStart routine and the card's master file is selected. The Card Management Utility displays the card's ATR as acknowledgment that the initialization was successful:

```
void CSmartShopperDlg::OnStart()
{
    char atr[20];
    CListBox *pBox;

    CDialog::OnInitDialog();

    pBox = (CListBox *)GetDlgItem(IDC_ATR);

    Initialize("PSCR_0");
    sprintf(atr, "%02x %02x %02x %02x ",
        scardIoResponse[0],scardIoResponse[1],
            scardIoResponse[2],scardIoResponse[3]);
    pBox->AddString(atr);
    SelectFile(0x3F00);
}
```

When the user selects the Enter PIN button, the Card Management Utility picks up the text in the PIN edit box, pads it with `0xFFs`, and uses it to call `VerifyCHV` on the card. The Card Management Utility displays the status code returned from the card; `9000` means the PIN was accepted. This will obviously be changed to something a little more intuitive in the next release of the Card Management Utility. Here is the Enter Pin button handler:

```
void CSmartShopperDlg::OnEnterpin()
{
    CListBox *pBox;
    BYTE pin[20];
    char sw[10];
    WORD length, i;

    pBox = (CListBox *)GetDlgItem(IDC_PINOK);

    *(WORD *)pin = sizeof(pin);
    length = (WORD)SendDlgItemMessage(IDC_PIN, EM_GETLINE, 0,
        (DWORD)(LPSTR)pin);
    for(i = length; i < 8; i++) pin[i] = 0xff;
    VerifyCHV(pin);
    sprintf(sw, "%04x", SW);
    pBox->ResetContent();
    pBox->AddString(sw);
}
```

When the user selects the Get Personal Profile button, the Card Management Utility selects the card file that contains the personal profile data, `0x0100`, and writes each record from this file into a listbox. The program uses the data word associated with each entry in the listbox to flag when an entry has been changed so it has to write back to the card only the records that are actually changed. Here is the Personal Profile button handler:

```
void CSmartShopperDlg::OnGetpp()
{
    CListBox *pBox;
    char item[25];
    BYTE i;

    pBox = (CListBox *)GetDlgItem(IDC_PP);
    SelectFile(0x0100);
    for(i = 0; i < 10; i++) {
        ReadRecord(i+1, 4, 24);
        memcpy(item, DATA, DATALENGTH);
        item[DATALENGTH] = '\0';
        pBox->InsertString(i,item);
        pBox->SetItemData(i, 0);
    }
}
```

Double-clicking a line in the personal profile data display copies the line to the edit window below the display, where it can be changed and updated. Here is the selected line handler:

```
int selitem;
char seltext[30];

void CSmartShopperDlg::OnDblclkPp()
{
    CListBox *pBox;
    CEdit *eBox;

    pBox = (CListBox *)GetDlgItem(IDC_PP);
    eBox = (CEdit *)GetDlgItem(IDC_EDITPP);
    selitem = pBox->GetCurSel();
    if(selitem != LB_ERR)
        pBox->GetText(selitem, seltext);
    eBox->SetWindowText(seltext);
}
```

After the customer finishes editing the double-clicked line, he selects the OK button next to it to move the new line back into the personal profile data display, where it overwrites the old line. Here is the done handler:

```
void CSmartShopperDlg::OnEditppdone()
{
    CListBox *pBox;
    CString item;
    CEdit *eBox;
    int i;

    pBox = (CListBox *)GetDlgItem(IDC_PP);
    eBox = (CEdit *)GetDlgItem(IDC_EDITPP);
    eBox->GetWindowText(item);
    i = item.GetLength();
    memcpy(seltext, (LPCTSTR)item, min(i,24));
    for(i = item.GetLength(); i < 24; i++) seltext[i] = ' ';
    seltext[24] = '\0';
    pBox->DeleteString(selitem);
    pBox->InsertString(selitem, seltext);
    pBox->SetItemData(selitem, 1); // mark item as changed
}
```

When all items that were to be updated have been updated, the user selects the Save Personal Profile button to write the changed records back to the Smart Shopper card. Here is the editing completion handler:

```
void CSmartShopperDlg::OnPpdone()
{
    CListBox *pBox;
    CString item;
```

```
    BYTE buffer[30];
    BYTE i;

    pBox = (CListBox *)GetDlgItem(IDC_PP);
    SelectFile(0x0100);
    for(i = 0; i < 10; i++) {
        if(pBox->GetItemData(i)) {
            pBox->GetText(i, item);
            memcpy(buffer, (LPCTSTR)item, 24);
            UpdateRecord(i+1, 4, 24, buffer, 0, 0);
        }
    }
    pBox->ResetContent();
}
```

Finally, the user can select the Examine Frequent Buyer Points button to view the
number of frequent buyer points accumulated in each Frequent Buyer Points program on the card. When this button is selected, the Card Management Utility first
reads the administration file, which contains the name of the participating merchant in each of the merchant directories. Then the Card Management Utility goes
into each of the installed merchant directories, reads the current record in the frequent buyer points file and displays what it finds next to the merchant's name.
Here's the code that displays frequent buyer point totals on the card:

```
void CSmartShopperDlg::OnGetfbp()
{
    CListBox *pBox;
    char s[100], *sp = s, *nsp;
    char t[100], v[20];
    int i, merchant, value;

    pBox = (CListBox *)GetDlgItem(IDC_FBPLIST);
    SelectFile(0x3F00);
    SelectFile(0x0F00);
    ReadBinary(0, 100);
    memcpy(s, DATA, 100);
    for(i = merchant = 0; i < 5; i++) {
        if((nsp = strchr(sp, '#')) == 0) {
            break;
        } else if(nsp-sp>1) {
            memset(t, ' ', 30);
            memcpy(t, sp, nsp-sp);
            SelectFile(4096*(i+1));
            if(SelectFile(4096*(i+1)+2)) {
                ReadRecord(0, 4, 6);
                value = DATA[2]<< 24 | DATA[3]<<16 |
                                DATA[4]<< 8 | DATA[5];
                sprintf(v, "%6d", value);
                memcpy(t+20, v, 7);
                t[27] = '\0';
                pBox->InsertString(merchant, t);
```

```
            pBox->SetItemData(merchant, 1000*(i+1));
            merchant++;
        }
    }
    sp = nsp+1;
    SelectFile(0x3F00);
    }
}
```

If you can think of ways to customize the Smart Shopper card Management Utility, get its code from the CD-ROM and start working!

Summary

This chapter describes a multimerchant smart card program, focusing primarily on the design of the smart card itself and the supporting application software. The intent is to show the design considerations that are characteristic of including a smart card in a consumer-oriented system. This chapter also discusses issues and concerns facing all smart card program stakeholders—card issuer, merchants and customers.

CHAPTER

THE FLEXCASH CARD: AN E-COMMERCE SMART CARD APPLICATION

I n this chapter, we will build an illustrative e-commerce application using Schlumberger's Cyberflex smart card and the Microsoft Windows PC/SC smart card infrastructure. The application consists of two programs: one for a Cyberflex smart card that's written in Java and one for a Windows PC that's written in C++.

The Cyberflex smart card runs programs written in any programming language that can be compiled into Java bytecodes (Java, Eiffel, C, C++, Ada, BASIC, and so on). A program on the Cyberflex card can communicate with applications on a personal computer or any other smart card terminal. As a result, the Cyberflex card can be programmed to do any of the following:

- Act like an existing smart card
- Act like more than one existing smart card
- Act like a wholly new smart card

In this chapter, we are going to program the Cyberflex smart card to do the jobs of three e-cash cards and a loyalty card.

The Microsoft PC/SC Windows smart card infrastructure, as discussed in Chapter 7, "Reader-Side Application Programming Interfaces," integrates smart cards and smart card readers in a very general way into the Microsoft Windows family of operating systems. Associated with almost any off-the-shelf smart card is a Windows library and application programming interface called a *PC/SC Smartcard Service Provider* (SSP) that lets your Windows program access the smart card. In this chapter, we write a Windows application in C++ and use the Cyberflex SSP.

The application programming interface on the Cyberflex card that the Cyberflex Java program uses and the application program interface on the PC that the C++ program uses both include GetMessage and SendMessage calls to communicate with the other party. On the card side, GetMessage and SendMessage are part of the Java Card 1.0 API and are implemented as Java native methods which access the ISO 7816 APDU communication capabilities of the underlying card operating system. On the reader side, the GetMessage and SendMesage functions in the Cyberflex SSP use the APDU communication capabilities on Microsoft's smart card API. Thus, our program on the Java Card and our program on the host PC are communicating with bona fide ISO 7816 APDUs.

The C++ program sends a message to the Java program using SendMessage and the Java program receives the message using GetMessage. The opposite happens, too—the Java program sends a message to the C++ program using SendMessage and the C++ program receives the message using GetMessage. As discussed in Chapter 4, "Smart Card Commands," the communication channel between the two programs is half-duplex, so they have to be synchronized to keep track of who is talking and who is listening. Figure 11.1 is a diagram of the relationship between the two parts of our e-commerce application program.

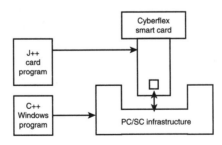

Figure 11.1. *FlexCash application program components.*

Because working on a specific problem is usually more interesting and informative than working on a general problem, we'll begin our application development by describing a mythical situation that our application is intended to address. After setting the stage, we'll discuss the design of a program for the smart card side of the application. Next, we'll discuss a simple protocol for moving value from the smart card to the PC program. Finally, we'll describe the Windows program.

An Example: E-Coins, E-Money, and E-Bucks

For the sake of this example, assume that there used to be three competing electronic cash cards: E-Coins, E-Money, and E-Bucks. Vending machines that accepted E-Coins cards wouldn't accept E-Money cards, and point-of-sale terminals that accepted E-Money cards wouldn't accept E-Bucks cards. Convincing people to switch from one kind of hard cash to three kinds of incompatible electronic cash was an uphill battle, and a losing battle at that. "Not only can I not see it, but I have to carry around three kinds of it" was the lament of the technology-weary man in the street. "Why can't those guys get their acts together?" Well, they finally did and all agreed to use ISO 7816-37 as the e-cash standard. But what to do about all those card-specific vending machines in the field while the world switched from the three proprietary systems to a unified and standardized e-cash system?

Smart Commerce Solutions, the imaginary folks that brought you the Smart Shopper card in Chapter 10, "The Smart Shopper Smart Card Program," came up with the FlexCash card. The FlexCash card carried all three electronic purses—E-Coins, E-Money and E-Bucks. When you stepped up to a vending machine that accepted only E-Coins cards, you took out your handy carry-along smart card reader, perhaps the Smart Commerce Solutions Smart Dock, and put your Smart Commerce Solutions FlexCash card into it. You pushed the E-Coins button on the Smart Dock, and the Smart Dock told the FlexCash card to start acting like an E-Coins card. You then stuck your FlexCash card in the E-Coins-centric vending machine. From the machine's point of view, the Smart Commerce Solutions Smart Card was just a plain old E-Coins card, so it debited the E-Coins purse and gave you your candy bar.

Besides the three electronic purses, the FlexCash card carried a frequent buyer points purse that accumulated Smart Commerce Solutions Smart Points based on your use of the card and independent of the three obsolete e-cash protocols you once used to spend your money.

Smart Commerce Solutions (hypothetically) used Schlumberger's Cyberflex card to conduct a field trial of the FlexCash card before committing its code to ROM.

The Design of the FlexCash Smart Card Program

The Java program on the FlexCash card consists of five modules: a monitor, three e-purses, and the frequent buyer points purse. Figure 11.2 shows how three modules connect to each other. The monitor communicates with the outside world and passes messages to the other four applications. This general architecture lets Smart Commerce Solutions quickly create smart cards that are multiapplication combinations of existing single-application cards.

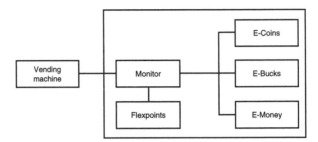

Figure 11.2. *Modules on the FlexCash card.*

Which application runs when the FlexCash card is activated depends on a flag stored in EEPROM called the *application flag*. If the application flag is 0, the FlexCash monitor application handles the interaction with the terminal. If the flag is some number *k* greater than 0, the monitor passes commands from the terminal to installed application *k* and sends the response generated by the application back to the terminal.

There is one special command that the monitor application handles no matter what the application flag is set to: SET APPLICATION FLAG. This lets the cardholder change from one application to another, regardless of what type of card the FlexCash card is currently acting as.

FlexCash Card Monitor Implementation

The FlexCash card monitor handles the following commands:

```
GET VALUES (E1 00 00 00 04)
DECREMENT LOYALTY by yy (E1 02 00 yy 00)
SET APPLICATION FLAG to xx (E1 04 00 xx 00)
```

The GET VALUES command returns the current values of the three purses and the loyalty points purse. It is used by the cardholder to find out how much value he or she has on his or her FlexCash card. The DECREMENT LOYALTY command subtracts a given value from the loyalty purse value. It is used to redeem loyalty purse points.

In passing commands from the terminal to one of the e-cash applications, the monitor recognizes the command for each application that represents spending money and increments the loyalty points purse corresponding to the amount of e-cash being deducted from the e-cash purse.

Listing 11.1 is the FlexCash monitor program, written in Java against the Java Card 1.0 API, which turns the general-purpose Cyberflex card into a FlexCash card. For the sake of clarity, this code assumes that all three purses have the same ATR, which would clearly not be the case if the cards really existed.

Listing 11.1. FlexCash code on the Java Card.

```
/*
** FlexCash Smart Card
*/

public class FlexCash {
    static final byte MONITOR = (byte) 0;
    static final byte ECOINS  = (byte) 1;
    static final byte EBUCKS  = (byte) 2;
    static final byte EMONEY  = (byte) 3;

    public static void main() {
        byte [] Array, Value, Ack;
        byte Application, Status;

        Array = new byte[8];

        // Send the first byte of the ATR
        Array[0] = (byte)0x3B;
        _OS.SendMessage(Array,(byte)0x01);

        // Reset the card during debugging
        //_OS.Execute((short)0,(byte)0);

        // Read in the purse values and the application flag
        _OS.SelectFile((short)0x7777);
        _OS.SelectFile((short)0x7701);
        Value = new byte[5];
        _OS.ReadBinaryFile((short)0,(byte)5, Value);
        Application = Value[4];

        // Fill the last part of the FlexCash Card ATR and send it
        if(Application == MONITOR) {
            Array[0] = (byte)0x32; Array[1] = (byte)0x15;
            Array[2] = (byte)0x00; Array[3] = (byte)0x49;
            Array[4] = (byte)0x10;
            _OS.SendMessage(Array,(byte)5);
        }

        /* Verify Key 0 */
        Array[0]=(byte)0xAD; Array[1]=(byte)0x9F;
            Array[2]=(byte)0x61; Array[3]=(byte)0xFE;
            Array[4]=(byte)0xFA; Array[5]=(byte)0x20;
        Array[6]=(byte)0xCE; Array[7]=(byte)0x63;
        _OS.VerifyKey((byte)0, Array, (byte)8);

        Ack = new byte[1];
```

continues

Listing 11.1. continued

```
while (true) {
  // Wait for a 5-byte command
  _OS.GetMessage(Array,(byte)0x05,(byte)0x00);
  Ack[0] = Array[1];

  // Trap the Set Application command
  if((Array[0] == (byte)0xE1) && (Array[1] == (byte)0x04)) {
    Value[4] = Ack[0] = Application = Array[3];
    Status = _OS.WriteBinaryFile((short)4, (byte)1, Ack);
    _OS.SendStatus(Status);
    continue;
  }

  // Pass the command to the active application
  switch(Application) {
  case MONITOR:
    if(Array[0] == (byte)0xE1) {
      if (Array[1] == (byte)0x02) { // Decrement Loyalty
        Value[0] = (byte)(Value[0]-Array[3]);
        Status =
                 _OS.WriteBinaryFile((short)0,(byte)1, Value);
        _OS.SendStatus(Status);
        continue;
      }
      if(Array[1] == (byte)0x00) { // Get Values
        _OS.SendMessage(Ack, (byte)1);
        _OS.SendMessage(Value, (byte)4);
        _OS.SendStatus((byte)0x00);
        continue;
      }
    } else { // Unknown Monitor command
        _OS.Execute((short)0,(byte)0);
        _OS.SendStatus((byte)0x90);
    }
  case ECOINS:
    // _OS.ECoins(Array);
    Value[1] = Ack[0] = (byte)(Value[1]-Array[3]);
    Status = _OS.WriteBinaryFile((short)1,(byte)1, Ack);
    _OS.SendStatus(Status);
    break;
  case EBUCKS:
    // _OS.EBucks(Array);
    Value[2] = Ack[0] = (byte)(Value[2]-Array[3]);
    Status = _OS.WriteBinaryFile((short)2,(byte)1, Ack);
    _OS.SendStatus(Status);
    break;
  case EMONEY:
    // _OS.EMoney(Array);
    Value[3] = Ack[0] = (byte)(Value[3]-Array[3]);
    Status = _OS.WriteBinaryFile((short)3,(byte)1, Ack);
```

```
                _OS.SendStatus(Status);
                break;
            default:
                _OS.SendStatus((byte)0x90);
                break;
            } // end of switch
        } // end of while
    } // end of main
} // end of class
```

The FlexCash Card Browser and Editor

The FlexCash card browser/editor program lets you toggle the Cyberflex card between being an off-the-shelf Version 1.0 Java Card and being a FlexCash card and, in turn, one of the three imaginary e-purse cards. When the card is set to be a Java Card, you can send the card any of the commands supported by the Java Card 1.0 bootstrap loader, including sending it a new Java program to run. The Java Card 1.0 bootstrap loader command set is implemented by code that is stored in and run from the ROM of the Cyberflex card. Since the point of the Cyberflex card is for you to provide the code that turns it into whatever card you like, there are just enough commands in the command set to let you get your code onto the card securely and kick it off. Here is the list of Java Card 1.0 boot loader commands:

- Select File
- Verify Key
- Create File
- Read Binary
- Write Binary
- Set Default ATR
- Execute Application

The only two of these commands that you haven't seen before in reading about ISO 7816 smart cards and that are in fact atypical for off-the-shelf smart cards are Set Default ATR and Execute Application.

The Set Default ATR command lets you set the ATR returned by the card when it is reset. The ATR from a smart card identifies what applications or application suites the smart card contains. It is used by smart card terminals and host application programs to determine if they are able to deal with the card. If you are going to make a Java Card behave like one or more existing smart cards, or if you are going to make it behave like a brand new smart card, you have to be able to set the ATR to meet the requirements of the terminals and host programs with which it will communicate. If you don't reset the ATR, the card simply identifies itself as an unprogrammed Java Card.

The `Execute Application` command instructs the card to execute your program when it is reset rather than the Java Card bootstrap loader. This is how your program takes over the Java Card and turns the card into the card defined by your program rather than a generic Java Card.

Rather than sporting a rich set of ISO 7816 commands, the Cyberflex card offers a rich set of function calls to your on-card program. It is your program that defines the commands to which the card responds. The commands supported by an off-the-shelf Java Card are only those needed to get your program onto the card and to let your program take charge of the card.

When the card is set to be a FlexCash card, you can send the card any of the commands supported by the FlexCash program in Listing 11.1. An executable version of this program is stored in and run from the transparent file 2222_{16}, which is in the master file of the Cyberflex card. After the program was loaded into this file, an `Execute Application` command was sent to the card telling it to "become" the program in file 2222_{16}; that is, to run the program in this file when it was reset. Once this is done, the card responds only to the commands implemented by the program and only the program can turn the card back into a Java Card.

Once we have turned the Java Card into a FlexCash card, we are ready to build the host application—a FlexCash card browser and editor that communicates with this card.

Figure 11.3 shows the human interface to the FlexCash card browser and editor and Listing 11.2 shows the MFC C++, which implements this human interface and the program behind it.

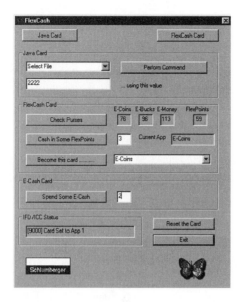

Figure 11.3. *The Human Interface of the FlexCash card browser and editor.*

Listing 11.2. The FlexCash card browser and editor program.

```
BYTE data[100], reply[100];
char string[100];
HRESULT hresult;
WORD wSW;
CString cString;
void UpdatePurseDisplay();
DWORD ApplicationFlag = 0;

void CFlexCashDlg::OnGetvalues()
{
    UpdatePurseDisplay();
}

void CFlexCashDlg::OnCashloyalty()
{
        DWORD amount;
        BYTE bpDecLoyalty[] = {0xE1, 0x02, 0x00, 0x03};

        m_LoyaltyCash.GetWindowText(cString);
        sscanf((LPCTSTR)cString, "%2d", &amount);
        bpDecLoyalty[3] = (BYTE)amount;

        hresult = SendCardMessage(bpDecLoyalty, (BYTE)0x04);
```

Listing 11.2. continued

```
                GetSW(&wSW);
                sprintf(string, "[%04x]", wSW);
                m_editScardStatus.SetWindowText(string);
    }

    void CFlexCashDlg::UpdatePurseDisplay()
    {
        BYTE bpGetValues[]    = {0xE1, 0x00, 0x00, 0x00};

        hresult =
            ExchangeCardMessage(bpGetValues, (BYTE)0x04, reply, 0x04);

        if (FAILED(hresult))
            throw (hresult);
        else {
            sprintf(string, "%3d", reply[1]);
            m_PStatus.SetWindowText(string);
            sprintf(string, "%3d", reply[2]);
            m_BStatus.SetWindowText(string);
            sprintf(string, "%3d", reply[3]);
            m_VStatus.SetWindowText(string);
            sprintf(string, "%3d", reply[0]);
            m_SStatus.SetWindowText(string);

            GetSW(&wSW);
            sprintf(string, "[%04x] %02x %02x %02x %02x %02x",
            wSW, reply[0], reply[1], reply[2], reply[3], reply[4]);
            m_editScardStatus.SetWindowText(string);
        }
    }

    void CFlexCashDlg::OnPerformcommand()
    {
        int iCurrentSelection = m_CommandList.GetCurSel();
        WORD wSW, wOffset;
        DWORD dwFileId, dwBytes;
        BYTE key[8] = {0xad, 0x9f, 0x61, 0xfe, 0xfa, 0x20, 0xce, 0x63};

        try {
            switch (iCurrentSelection) {
            case 0: // Create File
                m_editScardStatus.SetWindowText(
                        "[xxxx] Create File not supported.");
                break;
            case 1: // Execute
                hresult = Execute((WORD)0x2222, (BYTE)0x08);

                if (FAILED(hresult))
                    throw (hresult);
                else {
                    GetSW(&wSW);
```

```
                sprintf(string,
                        "[%04x] Boot File Set to 0x2222.");
                m_editScardStatus.SetWindowText(string);
            }
            break;
        case 2: // Exchange Message
            m_CommandArg.GetWindowText(cString);
            sscanf((LPCTSTR)cString, "%2x%2x%2x%2x%2x",
                &data[0],&data[1],&data[2],&data[3],&data[4]);

            hresult =
                    ExchangeCardMessage(data, (BYTE)0x04, reply, data[4]);

            if (FAILED(hresult))
                throw (hresult);
            else {
                GetSW(&wSW);
                sprintf(string, "[%04x] %02x %02x %02x %02x",
                wSW, reply[0], reply[1], reply[2], reply[3]);
                m_editScardStatus.SetWindowText(string);
            }

            break;
        case 3: // Read Binary
            m_CommandArg.GetWindowText(cString);
            sscanf((LPCTSTR)cString, "%d", &wOffset);
            hresult = ReadBinary(reply, wOffset, (BYTE)0x08);

            if (FAILED(hresult))
                throw (hresult);
            else {
                GetSW(&wSW);
                sprintf(string, "[%04x] %02x %02x %02x
%02x %02x %02x %02x %02x",
                wSW, reply[0], reply[1], reply[2], reply[3],
                    reply[4], reply[5], reply[6], reply[7]);
                m_editScardStatus.SetWindowText(string);
            }
            break;
        case 4: // Select Master File (3F00)

            hresult = SelectFile((WORD)0x3F00);

            if (FAILED(hresult))
                throw (hresult);
            else {
                GetSW(&wSW);
                sprintf(string,
                        "[%04x] Master File Selected", wSW);
                m_editScardStatus.SetWindowText(string);
            }
            break;
```

continues

Listing 11.2. continued

```
        case 5: // Select File
            m_CommandArg.GetWindowText(cString);
            sscanf((LPCTSTR)cString, "%x", &dwFileId);
            hresult = SelectFile((WORD)dwFileId);

            if (FAILED(hresult))
                throw (hresult);
            else {
                GetSW(&wSW);
                sprintf(string, "[%04x] File Selected", wSW);
                m_editScardStatus.SetWindowText(string);
            }
            break;
        case 6: // Set Default ATR
            m_CommandArg.GetWindowText(cString);
            dwBytes = cString.GetLength();
            CopyMemory(data, (LPCSTR)cString, dwBytes);

            hresult = SetDefaultATR(data, (BYTE)dwBytes);

            if (FAILED(hresult))
                throw (hresult);
            else {
                GetSW(&wSW);
                sprintf(string, "[%04x] ATR Set.", wSW);
                m_editScardStatus.SetWindowText(string);
            }
            break;
        case 7: // Verify Key

            hresult = VerifyKey((BYTE)0x00, key, (BYTE)0x08);

            if (FAILED(hresult))
                throw (hresult);
            else {
                GetSW(&wSW);
                sprintf(string, "[%04x] Key Verified.", wSW);
                m_editScardStatus.SetWindowText(string);
            }
            break;
        case 8: // Write Binary
            m_CommandArg.GetWindowText(cString);
            dwBytes = sscanf((LPCTSTR)cString,
                        "%2x%2x%2x%2x%2x%2x%2x%2x",
                &data[0],&data[1],&data[2],&data[3],
                &data[4],&data[5],&data[6],&data[7]);

            hresult = WriteBinary(data, (WORD)0, (BYTE)dwBytes);

            if (FAILED(hresult))
                throw (hresult);
```

```
            else {
                GetSW(&wSW);
                sprintf(string, "[%04x] Data Written.", wSW);
                m_editScardStatus.SetWindowText(string);
            }
            break;
        default:
            break;
        };
    }

    catch (HRESULT) {
        GetSW(&wSW);
        sprintf(string, "[%02x/%08x]", wSW, hresult);
        FormatMessage(FORMAT_MESSAGE_FROM_SYSTEM, NULL, hresult,
                MAKELANGID(LANG_NEUTRAL, SUBLANG_DEFAULT),
                &string[strlen(string)], 100, NULL);
        m_editScardStatus.SetWindowText(string);
    }
}

void CFlexCashDlg::OnBecomeFlexCash()
{
    BYTE key[8] = {0xad, 0x9f, 0x61, 0xfe, 0xfa, 0x20, 0xce, 0x63};

        DO(VerifyKey((BYTE)0x00, key, (BYTE)0x08))
    && DO(Execute((WORD)0x2222, (BYTE)0x08))
    && DO(ResetCard());

    sprintf(string, "[%08x] FlexCash Card Activated.", hresult);
    m_editScardStatus.SetWindowText(string);
}

void CFlexCashDlg::OnBecomeJavaCard()
{
    BYTE bpBecomeJavaCard[] = {0xE0, 0x00, 0x00, 0x00};

        DO(SendCardMessage(bpBecomeJavaCard, (BYTE)0x04))
    && DO(ResetCard());

    sprintf(string, "[%08x] Java Card Activated.", hresult);
    m_editScardStatus.SetWindowText(string);
}

void CFlexCashDlg::OnReset()
{
    hresult = ResetCard();

    sprintf(string, "[%08x] Card Reset.", hresult);
    m_editScardStatus.SetWindowText(string);
}
```

Listing 11.2. continued

```
void CFlexCashDlg::OnChangecard()
{
    BYTE bpSetApplication[] = {0xE1, 0x04, 0x00, 0x00};
    char *ecash[] = {"Monitor", "E-Coins", "E-Bucks", "E-Money"};

    bpSetApplication[3] = (BYTE)m_CardType.GetCurSel();
    hresult = SendCardMessage(bpSetApplication, (BYTE)0x04);

    m_CurrentApp.SetWindowText(ecash[bpSetApplication[3]]);

    GetSW(&wSW);
    sprintf(string, "[%04x] Card Set to App %d", wSW,
            (DWORD)bpSetApplication[3]);
    m_editScardStatus.SetWindowText(string);
}

void CFlexCashDlg::OnSpendecash()
{
        DWORD amount;
        BYTE bpSpendECash[] = {0xE2, 0x00, 0x00, 0x00};

        m_ECashAmount.GetWindowText(cString);
        sscanf((LPCTSTR)cString, "%2d", &amount);
        bpSpendECash[2] = (BYTE)ApplicationFlag;
        bpSpendECash[3] = (BYTE)amount;

        hresult = SendCardMessage(bpSpendECash, (BYTE)0x04);

        GetSW(&wSW);
        sprintf(string, "[%04x]", wSW);
        m_editScardStatus.SetWindowText(string);
}
```

The E-Bucks E-cash Protocol and Implementation

In our storybook world, the E-Coins and E-Money e-cash purses were sufficiently complicated that it was decided to simply put the object code implementing these purses onto the FlexCash card as soft masks. They appear as Java native function calls to the monitor. Since the code implementing these purses is exactly the same as the code on the standalone E-Coins and E-Money cards, when the application flag is set to E-Coins, the card acts like an E-Coins card. When the application flag is set to E-Money, the card acts like an E-Money card.

As it turns out, the E-Bucks e-purse was implemented for a proprietary microcontroller chip, so the object code for the E-Bucks purse couldn't be soft masked onto the FlexCash card. Fortunately, the E-Bucks e-cash protocol was very simple and could be implemented in Java. The next section describes the simple E-Bucks e-cash protocol and the Java code that implements it.

The E-Bucks E-cash Protocol

One of the simplest e-cash protocols between a merchant's till and a customer's smart card is a two-stage commit protocol, which goes like this:

1. Debit

   Till: "Please debit the customer's purse by *X* dollars."

   Card: "Okay"

2. Credit

   Till: "Please give me permission to credit the merchant by *X* dollars."

   Card: "Okay"

While this is a very simple value-transfer protocol, if it is combined with some common-sense security measures it could actually be used. Here are some of the security considerations that have to be taken into account.

First, notice that the Okay that says the customer's card has been debited does not also mean that it is okay to credit the merchant's till. The card must give explicit authorization to perform the credit. This is a characteristic of many e-cash protocols. Each message means exactly one thing, and the protocol proceeds exactly one step at a time.

Second, notice that the merchant's till is credited only after the card has been debited. A customer who is debited and receives nothing because a merchant wasn't credited will complain, whereas a merchant who receives a credit without a customer being debited will not. Said a different way, if a mistake is to be made, it is better from the card issuer's point of view for money to be destroyed than for money to be minted.

Third, while you don't see it here, each message from the till to the card and from the card to the till is encrypted with a shared secret key. This is to counter man-in-the-middle attacks. If all an attacker posing as a valid smart card had to do was say "okay" in plain, unencrypted text, merchandise could be purchased without a valid card being debited. Furthermore, as an extra security measure, the shared key used to debit the card could be different from the shared key used to credit the till.

Fourth, each encrypted communication from the till to the card includes a transaction number, together with a challenge generated by the card. Similarly, each encrypted communication from the card to the till includes a transaction number along with a challenge generated by the till, including a transaction number and a challenge in the message counter's replay attacks. A *replay attack* grabs an encrypted message off the communication channel and, without having to decrypt it, simply sends it to the recipient again later. The recipient might think it is a valid message because it is encrypted with the proper key. For example, with unique message identifiers such as the transaction number and the random challenge, an attacker could simply grab an encrypted okay message going from the card to the till and play it back to the till whenever it asked for something.

The transaction number and the challenge are explicit indicators of whether a message is a credit or a debit message. They ensure that a debit message from the till to the card will not be turned around and sent back to the till as a credit message or vice versa.

The full, simple-but-effective E-Bucks e-cash protocol looks like this:

Till: "Receive a cardholder request to debit a proffered card. Ask the card for a transaction number and a challenge."

Card: "Generate transaction number and challenge. Send them to the till. Remember the number and the challenge."

Till: "Generate encrypted message containing amount of debit, the transaction number, the challenge, and the debit flag. Send this message to the card."

Card: "Check whether the debit message is authentic by finding the previously sent challenge and transaction number along with the debit message indicator when the message is decrypted. If all is well, debit the customer's card, remember the amount, and set a flag that a request to authorize a credit is outstanding. Send an encrypted Okay back to the till."

Till: "Generate a transaction number and a challenge. Send them to the card with a request to authorize a credit."

Card: "Generate an encrypted message containing the remembered amount of the transaction, the transaction number, the challenge, and the credit message indicator. Clear the credit-request-outstanding flag and send the message to the till."

Till: "Check whether the credit message is authentic by finding the previously sent challenge and transaction number along with the credit message when the message is decrypted. If all is well, credit the merchant's till."

This simple e-cash protocol is implemented on the E-Bucks card using the commands shown in the following list of E-Bucks and e-cash commands:

REQUEST DEBIT

Bytecodes: $E2_{16}$ 00_{16} 00_{16} 00_{16} 00_{16}

Parameters: None

Data: None

Reply: Status plus 2 bytes, *xx yy*, where *xx* is the E-Bucks card's current transaction number and *yy* is the E-Bucks card's challenge

MAKE DEBIT

Bytecodes: $E2_{16}$ 02_{16} 00_{16} 00_{16} 08_{16} zz_{16} zz_{16} zz_{16} zz_{16}

Parameters: None

Data: The encryption of the amount, the E-Bucks transaction number, the E-Bucks challenge, and the debit flag

Reply: Status

REQUEST CREDIT

Bytecodes: $E2_{16}$ 04_{16} xx_{16} yy_{16} 00_{16}

Parameters: The till's transaction number, xx, and the till's challenge, yy.

Data: None

Reply: Status plus 8 bytes, zz_{16} zz_{16} zz_{16} zz_{16}, which is the encryption of the till's transaction number, the till's challenge, the amount of the transaction from the previous MAKE DEBIT command, and the credit flag.

The E-Bucks Card-Side Code

Listing 11.3 is the new and improved E-Bucks section of the FlexCash smart card program that implements the E-Bucks e-cash protocol. For the sake of clarity, the Java code doesn't use any encryption.

Listing 11.3. E-Bucks e-cash card-side code.

```
case EBUCKS:
    if(Array[0] == (byte)0xE2) {
        // Get EBucks Transaction Number
        _OS.ReadBinaryFile((short)5, (byte)1, EBucks);
        if(Array[1] == (byte)0x02) { // Request Debit
            EBucks[0]++;
            EBucks[1] = CHALLENGE;
            Status =
```

Listing 11.3. continued

```
                        _OS.WriteBinaryFile((short)5, (byte)1, EBucks);
            _OS.SendMessage(Ack, (byte)1);
            _OS.SendMessage(EBucks, (byte)2);
            _OS.SendStatus(Status);
            continue;
        }
        if(Array[1] == (byte)0x04) { // Make Debit
          _OS.GetMessage(Array, (byte) 0x04, Ack[0]);
          if(Array[0] == EBucks[0] && // Transaction Number
             Array[1] == CHALLENGE &&
             Array[2] == DEBITFLAG) {
              EBucks[1] = CREDITFLAG;
              EBucks[2] = Array[3]; // Amount
              Value[2] = Ack[0] = (byte)(Value[2]-Array[3]);
              Status = _OS.WriteBinaryFile((short)2,(byte)1, Ack);
              _OS.SendStatus(Status);
              continue;
          } else {
              _OS.SendStatus((byte)0x30);
              continue;
          }
        }
        if(Array[1] == (byte)0x06) { // Request Credit
          if(EBucks[1] == CREDITFLAG) {
              Array[0] = Array[2];  // Transaction Number
              Array[1] = Array[3];  // Challenge
              Array[2] = EBucks[2]; // Amount
              Array[3] = CREDITFLAG;
              EBucks[1] = 0;
              _OS.SendMessage(Ack, (byte)1);
              _OS.SendMessage(Array, (byte)4);
              _OS.SendStatus((byte)0x00);
              continue;
          } else {
              _OS.SendStatus((byte)0x30);
              continue;
          }
        }
    } else { // Unknown E-Bucks command
    _OS.Execute((short)0,(byte)0);
    _OS.SendStatus((byte)0x95);
    continue;
    }
break;
```

The E-Bucks Reader-Side Code

To complete our story, Listing 11.4 is a Windows PC program that might run on a vending machine that accepts E-Bucks e-cash cards.

Listing 11.4. E-Bucks e-cash host-side code.

```
#define DEBITFLAG 1

void CSmartCashDlg::OnSpendebucks()
{
    DWORD dwAmount;

    BYTE bpRequestDebit[]  = {0xE2, 0x02, 0x00, 0x00};
    BYTE bpMakeDebit[]     = {0xE2, 0x04, 0x00, 0x00,
                              0x00, 0x00, 0x00, 0x00};
    BYTE bpRequestCredit[] = {0xE2, 0x06, 0x00, 0x00, 0x00};

    m_EBucks.GetWindowText(cString);
    sscanf((LPCTSTR)cString, "%2d", &dwAmount);

    hresult =
        ExchangeCardMessage(bpRequestDebit, (BYTE)0x04, bpReply, 0x02);
    GetSW(&wSW);

    bpMakeDebit[4] = bpReply[0];
    bpMakeDebit[5] = bpReply[1];
    bpMakeDebit[6] = DEBITFLAG;
    bpMakeDebit[7] = (BYTE)dwAmount;
    hresult = SendCardMessage(bpMakeDebit, (BYTE)0x08);
    GetSW(&wSW);

    hresult =
        ExchangeCardMessage(bpRequestCredit, (BYTE)0x04, bpReply, 0x04);

    if (FAILED(hresult))
        throw (hresult);
    else {
        GetSW(&wSW);
        sprintf(string, "[%04x] %02x %02x %02x %02x",
            wSW, bpReply[0], bpReply[1], bpReply[2], bpReply[3]);
            m_editScardStatus.SetWindowText(string);
    }
}
```

Summary

In this chapter, we consider a simple e-commerce smart card application that illustrates writing code to run on a smart card together with some basic security considerations in moving value from the card to the host. Be assured that the protocol used by real e-cash cards such as VisaCash, Mondex, and Proton is much more complicated than this one. But if you are just running a frequent buyer points program for Joe's Fish Store, then the preceding e-cash protocol would probably provide sufficient security to move Pisces Points between Joe's cash register and the smart card.

PART IV

APPENDIXES

APPENDIX

THE ISO 7816-4 COMMAND SET

The ISO 7816-4 standard defines a set of inter-industry commands that are meant to be included on smart cards. These commands are included in whole, or in part, on many smart cards available on the market today. This appendix gives an overview of the individual commands in this set, a summary of the status/error messages that may be returned, which command application protocol data unit (APDU) is sent to the APDU processor on a smart card, and, where feasible, an example APDU for this command.

APDU Structures

The APDU structures in which these commands would be transported is reviewed in Chapter 4, "Smart Card Commands." The constituent elements of the structures are

- CLA—The 1-byte designation of a family of commands.

- INS—The 1-byte designation of a specific command in this family.

- P1—A 1-byte parameter passed along as part of the [CLA,INS] command that elaborates on the exact meaning of the command; a command modifier.

- P2—A 1-byte parameter passed along as part of the [CLA,INS] command that elaborates on the exact meaning of the command; a command modifier.

- Lc field—a field that specifies the length of the data field (which follows). For essentially all existing cards, the size of this field is 1 byte, so it can define a data field length up to 256 bytes. However, it should be noted that a mechanism is defined within the ISO 7816-4 standard through which a card can define an extended address space which would allow longer fields to be specified. We will limit our discussion to the typical case where the Lc field is 1 byte in length.

- Data field—a string of bytes whose length is specified by the Lc field. These bytes are conveyed via the APDU to the card's APDU processor.

- Le field—a field that specifies the length of the body of the response APDU (to this command); this number of bytes is returned by the card's APDU processor on successful completion of the command. As with the Lc field, a card can define an extended addressing facility; however, we will limit our current review to 1 byte Le fields.

Security Status

Access to files through the commands described in this appendix is limited by a requirement that the security status of the card satisfy the security attributes defined for the files being accessed. The security status of the card is typically established through the successful execution of commands defined in the section "Security" later in this appendix.

Security attributes ascribed to a file can require the reader-side application component to present knowledge of a password (known by the card), providing knowledge of a key that's shared with the card, or through the use of secure messaging.

File System

The file system commands comprise a set of commands through which a file system on the card can be accessed by a reader-side application. It is interesting to note that two file operations that you would typically find associated with a file system are not present within this family; that is, a file create command and a file delete command. Commands such as these are found on many smart cards (such as the Multiflex card); however, their semantics are not defined through the ISO 7816-4 Standard.

Read Binary

Description

This command causes a portion of the selected file to be read and passed back through the response message. The file segment to be read is specified through a byte offset from the beginning of the file and a byte count of the number of bytes to be read. This command uses a Case 2 APDU structure; that is, the APDU includes a complete header along with an Le field which specifies the number of bytes to be returned.

When this command is executed, a `Select File` command will typically already have been issued to select the file to actually read. However, the `Select File` command may have pointed at a DF, which contains the EF to be read by this command. In that case, the P1 parameter can be used to convey a short EF identifier (that is, a 5-bit value that uniquely specifies an EF within a DF).

Command Application Protocol Data Unit

CLA	INS	Parameter 1	Parameter 2	Parameter 3
CO_{16}	BO_{16}	Short EF identifier	Offset of first byte read	Number of bytes to be read

Data Field 1	Data Field 2
N/A	N/A

Response Application Protocol Data Unit

Response

The number of bytes that were requested to be read, followed by the 2-byte status

Example of Use

APDU	Interpretation
$CO_{16}\ BO_{16}\ 00_{16}\ 00_{16}\ 10_{16}$	Read 16 bytes from the currently selected transparent file starting with the first byte in the file

Error Codes (As Specified by ISO/IEC 7816-4: 1995(E))

Status	Meaning	Status	Meaning
6281_{16}	Return data error	6282_{16}	Premature EOF
6700_{16}	Incorrect field	6981_{16}	Bad command

continues

273

Status	Meaning	Status	Meaning
6982_{16}	Invalid security status	6986_{16}	EF not selected
$6A81_{16}$	Invalid function	$6A82_{16}$	File missing
$6B00_{16}$	Invalid parameters	$6CXX_{16}$	Incorrect Le field

Write Binary

Description

This command provides for setting the values of specified bytes of the selected file. Depending on the file's attributes, the write operation may result in ANDing or ORing the bytes specified in the command with the values already in the file.

Command Application Protocol Data Unit

CLA	INS	Parameter 1	Parameter 2	Parameter 3
CO_{16}	DO_{16}	Short EF identifier	Offset of first byte written	Number of bytes to be written

Data Field 1	Data Field 2
String of bytes to be written	Empty

Response Application Protocol Data Unit

Response
2-byte status

Example of Use

APDU	Interpretation
$CO_{16}\ DO_{16}\ 01_{16}\ 01_{16}\ 01_{16}\ FF_{16}$	Select EF file 1 (by short identifier) within the currently selected DF, and then write all 1s in the second byte of the file, assuming that file attributes are correct

Error Codes (As Specified by ISO/IEC 7816-4: 1995(E))

Status	Meaning	Status	Meaning
$63CX_{16}$	Success with retries	6581_{16}	Invalid write
6700_{16}	Invalid Le field	6981_{16}	Bad command

Status	Meaning	Status	Meaning
6982_{16}	Invalid security status	6986_{16}	EF not selected
$6A81_{16}$	Invalid function	$6A82_{16}$	File missing
$6B00_{16}$	Invalid parameters		

Update Binary

Description

This command provides for setting the values of specified bytes of the selected file. This command functions essentially like a file `write` command. The resulting values of the file are those indicated in the command.

Command Application Protocol Data Unit

CLA	INS	Parameter 1	Parameter 2	Parameter 3
$C0_{16}$	$D6_{16}$	Short EF identifier	Offset of first byte written	Number of bytes to be written

Data Field 1	Data Field 2
Byte string to be written	Empty

Response Application Protocol Data Unit

Response

2-byte status

Example of Use

APDU	Interpretation
$C0_{16}\ D6_{16}\ 01_{16}\ 01_{16}\ 01_{16}\ FF_{16}$	Select EF file 1 (by short identifier) within the currently selected DF, and then write all 1s in the second byte of the file

Error Codes (As Specified by ISO/IEC 7816-4: 1995(E))

Status	Meaning	Status	Meaning
$63CX_{16}$	Success with retries	6581_{16}	Invalid write
6700_{16}	Incorrect Le field	6981_{16}	Bad command

continues

275

Status	Meaning	Status	Meaning
6982_{16}	Invalid security status	6986_{16}	EF not selected
$6A81_{16}$	Invalid function	$6A82_{16}$	File missing
$6B00_{16}$	Invalid parameters		

Erase Binary

Description

This command results in the setting of specified bytes of the selected file to a logical erased state. In general, this state is equivalent to a 0 value. The command works by spacing across the offset number of bytes and starting to erase. It then terminates on the byte specified by the parameters, or at the end of the file. So, it's possible to erase a segment of bytes within a file if that is desired.

Command Application Protocol Data Unit

CLA	INS	Parameter 1	Parameter 2	Parameter 3
$C0_{16}$	$0E_{16}$	Short EF identifier	Offset of first byte erased	If not zero this is length of data field 1

Data Field 1	Data Field 2
If present, this is the offset of the first byte not erased; this offset must be greater than the offset in parameter 2	Empty

Response Application Protocol Data Unit

Response
2-byte status

Example of Use

APDU	Interpretation
$C0_{16}\ 0E_{16}\ 01_{16}\ 01_{16}\ 01_{16}\ 06_{16}$	Select EF file 1 (by short identifier) within the currently selected DF, and then erase the second byte of the file through the sixth byte of the file

Error Codes (As Specified by ISO/IEC 7816-4: 1995(E))

Status	Meaning	Status	Meaning
$63CX_{16}$	Success with retries	6581_{16}	Invalid write

Status	Meaning	Status	Meaning
6700_{16}	Incorrect Le field	6981_{16}	Bad command
6982_{16}	Invalid security status	6986_{16}	EF not selected
$6A81_{16}$	Invalid function	$6A82_{16}$	File missing
$6B00_{16}$	Invalid parameters		

Read Record

Description

This command provides for reading one to several records of a file that has an internal record structure.

Command Application Protocol Data Unit

CLA	INS	Parameter 1	Parameter 2	Parameter 3
$C0_{16}$	$B2_{16}$	Index of the record to be read (01_{16}, 02_{16}, ..., FF_{16}) OR 00_{16} if the current record is to be read	Selection of record to be read: 00 first record; 01 last record; 02 next record; 03 previous record; 04 current record; if index is 0 or index record if it isn't	The number of bytes to be read from the record identified by P1 and P2; must be equal to the length of the record in the file

Data Field 1	Data Field 2
Empty	Empty

Response Application Protocol Data Unit

Response

The number of bytes in the record if the command is successful followed by the usual 2-byte status

Example of Use

APDU	Interpretation
$C0_{16}$ $B2_{16}$ 06_{16} 04_{16} 14_{16}	The records in the selected fixed-length record file are 20 bytes long; this command reads the sixth record in the file

Error Codes (As Specified by ISO/IEC 7816-4: 1995(E))

Status	Meaning	Status	Meaning
6281_{16}	Return data error	6282_{16}	Premature EOF
6700_{16}	Incorrect Le field	6981_{16}	Bad command
6982_{16}	Invalid security status	$6A81_{16}$	Invalid function
$6A82_{16}$	File missing	$6A83_{16}$	Missing record
$6CXX_{16}$	Wrong Le length		

Write Record

Description

This command provides for writing one record into a file that has an internal record structure. Depending on the file's attributes, the write operation may result in ANDing or ORing the bytes specified in the command with the values already in the file. For the example, we'll assume that the attributes are set for a one-time-write operation; that is, this will be the initial writing of data into this record of the file.

Command Application Protocol Data Unit

CLA	INS	Parameter 1	Parameter 2	Parameter 3
$C0_{16}$	$D2_{16}$	Index of the record to be written to (01_{16}, 02_{16}, ..., FF16) OR 0 if the current record is to be overwritten	Selection of record to be affected: 00 first record; 01 last record; 02 next record; 03 previous record; 04 current record; if index is 0 or index record if it isn't	The number of bytes to be merged into the record identified by Parameter 1 and Parameter 2; must be equal to the length of the record in the file

Data Field 1	Data Field 2
The data bytes to be written into the record identified by P1 and P2	Empty

Response Application Protocol Data Unit

Response
2-byte status

Example of Use

APDU	Interpretation
$C0_{16}$ $D2_{16}$ 06_{16} 04_{16} 14_{16} 53_{16} 61_{16} $6C_{16}$ $6C_{16}$ 79_{16} 20_{16} 47_{16} 72_{16} 65_{16} 65_{16} $6E_{16}$ 00_{16} 00_{16} 00_{16} 00_{16} 00_{16} 00_{16} 00_{16} 00_{16} 00_{16}	The records in the selected fixed-length record file are 20 bytes long; this command writes Sally Green into the sixth record in this file

Error Codes (As Specified by ISO/IEC 7816-4: 1995(E))

Status	Meaning	Status	Meaning
$63CX_{16}$	Success with retries	6581_{16}	Invalid write
6700_{16}	Incorrect Le field	6981_{16}	Bad command
6982_{16}	Invalid security status	6986_{16}	EF not selected
$6A81_{16}$	Invalid function	$6A82_{16}$	File missing
$6A83_{16}$	Missing record	$6A84_{16}$	Insufficient file space
$6A85_{16}$	Invalid TLV		

Append Record

Description

This command provides for either the appending of a record at the end of an EF with a linear structure or the writing of the first record of a cyclic, structured file.

Command Application Protocol Data Unit

CLA	INS	Parameter 1	Parameter 2	Parameter 3
$C0_{16}$	$E2_{16}$	00	Short EF file identifier	The number of bytes in the append record

Data Field 1	Data Field 2
Contents of the append record	Empty

Response Application Protocol Data Unit

Response

2-byte status

Example of Use

APDU	Interpretation
CO_{16} $D2_{16}$ 00_{16} 00_{16} 14_{16} 53_{16} 61_{16} $6C_{16}$ $6C_{16}$ 79_{16} 20_{16} 47_{16} 72_{16} 65_{16} 65_{16} $6E_{16}$ 00_{16} 00_{16} 00_{16} 00_{16} 00_{16} 00_{16} 00_{16} 00_{16} 00_{16}	The records in the selected fixed-length record file are 20 bytes long; this command appends Sally Green onto the end of the EF

Error Codes (As Specified by ISO/IEC 7816-4: 1995(E))

Status	Meaning	Status	Meaning
$63CX_{16}$	Success with retries	6581_{16}	Invalid write
6700_{16}	Incorrect Le field	6981_{16}	Bad command
6982_{16}	Invalid security status	6986_{16}	EF not selected
$6A81_{16}$	Invalid function	$6A82_{16}$	File missing
$6A83_{16}$	Missing record	$6A84_{16}$	Insufficient file space
$6A85_{16}$	Invalid TLV		

Update Record

Description

This command provides for updating (writing) a specific set of bytes in a specified record of a file.

Command Application Protocol Data Unit

CLA	INS	Parameter 1	Parameter 2	Parameter 3
CO_{16}	DC_{16}	Index of the record to be overwritten (01_{16}, 02_{16}, ..., FF_{16}) OR 0 if the current record is to be overwritten	Selection of record to be overwritten: 00 first record; 01 last record; 02 next record; 03 previous record; 04 current record; if index is 0 or index record if it isn't	The number of bytes to be written into the record identified by Parameter 1 and Parameter 2; must be equal to the length of the record in the file

Data Field 1	Data Field 2
The data bytes to be written into the record identified by P1 and P2	Empty

Response Application Protocol Data Unit

Response

2-byte status

Example of Use

APDU	Interpretation
CO_{16} DC_{16} 06_{16} 04_{16} 14_{16} 53_{16} 61_{16} $6C_{16}$ $6C_{16}$ 79_{16} 20_{16} 47_{16} 72_{16} 65_{16} 65_{16} $6E_{16}$ 00_{16} 00_{16} 00_{16} 00_{16} 00_{16} 00_{16} 00_{16} 00_{16} 00_{16}	The records in the selected fixed-length record file are 20 bytes long; this command writes Sally Green into the sixth record in this file

Error Codes (As Specified by ISO/IEC 7816-4: 1995(E))

Status	Meaning	Status	Meaning
$63CX_{16}$	Success with retries	6581_{16}	Invalid write
6700_{16}	Incorrect Le field	6981_{16}	Bad command
6982_{16}	Invalid security status	6986_{16}	EF not selected
$6A81_{16}$	Invalid function	$6A82_{16}$	File missing
$6A83_{16}$	Missing record	$6A84_{16}$	Insufficient file space
$6A85_{16}$	Invalid TLV		

Get Data

Description

This command provides for the reading of one primitive data object. In the context of this class of commands, a data object is one of two types of data structures: a BER-TLV structure or a SIMPLE-TLV structure. TLV means a *tag, length,* and *value* structure (that is, a structure where a tag field gives an identity to the structure, a length field gives a size [in bytes], and a value field contains the piece of information that is the reason for the existence of the structure).

In a SIMPLE-TLV data object, the tag field is a single byte containing a number that identifies the data object, the length field consists of 1 or 3 bytes in length (if 1 byte the length is 0 to 254 bytes, if three the length is 0 to 64K bytes), and the value field is a string of bytes of length given by the length field. A BER-TLV structure is defined by the ISO/IEC 8825 standard.

Command Application Protocol Data Unit

CLA	INS	Parameter 1	Parameter 2	Parameter 3
CO_{16}	CA_{16}	Data object	Identifier	Size of response Le

Data Field 1	Data Field 2
Empty	Empty

Response Application Protocol Data Unit

Response

Le bytes of data followed by 2-byte status

Example of Use

APDU	Interpretation
CO_{16} CA_{16} 02_{16} 01_{16} 14_{16}	This command retrieves up to 20 bytes of the value of a SIMPLE-TLV structure with ID = 01

Error Codes (As Specified by ISO/IEC 7816-4: 1995(E))

Status	Meaning	Status	Meaning
6281_{16}	Returned data error	6700_{16}	Incorrect Le field
6982_{16}	Invalid security status	6985_{16}	Invalid conditions
$6A81_{16}$	Invalid function	$6A88_{16}$	Missing data object
$6CXX_{16}$	Wrong Le length		

Put Data

Description

This command provides for the writing of one data object or of several data objects that have been packed into one constructed data object.

Command Application Protocol Data Unit

CLA	INS	Parameter 1	Parameter 2	Parameter 3
CO_{16}	DA_{16}	Data object	Identifier	Length of data field

Data Field 1	Data Field 2
Data to be written	Empty

Response Application Protocol Data Unit

Response

2-byte status

Example of Use

APDU	Interpretation
CO_{16} DA_{16} 02_{16} 01_{16} 01_{16} FF_{16}	This command stores a 1-byte value of all ones in the SIMPLE-TLV structure with ID = 01

Error Codes (As Specified by ISO/IEC 7816-4: 1995(E))

Status	Meaning	Status	Meaning
$63CX_{16}$	Success with retries	6581_{16}	Invalid write
6700_{16}	Incorrect Le field	6982_{16}	Invalid security status
6985_{16}	Invalid conditions	$6A80_{16}$	Incorrect data parameters
$6A81_{16}$	Invalid function	$6A84_{16}$	Insufficient file space
$6A85_{16}$	Invalid TLV		

Select File

Description

This command establishes a specific file which will then be the target of any subsequent file operation commands.

Command Application Protocol Data Unit

CLA	INS	Parameter 1	Parameter 2	Parameter 3
CO_{16}	$A4_{16}$	00_{16}	00_{16}	02_{16}

Data Field 1	Data Field 2
2-byte file identifier	N/A

Response Application Protocol Data Unit

Response

2-byte status; if the high-order byte of the status word is 61_{16}, then the low-order byte is the number of bytes of file description data that can be retrieved with a subsequent `Get Response` command

Example of Use

APDU	Interpretation
$C0_{16}$ $A4_{16}$ 00_{16} 00_{16} 02_{16} $3F_{16}$ 00_{16}	The master file becomes the currently selected directory

Error Codes (As Specified by ISO/IEC 7816-4: 1995(E))

Status	Meaning	Status	Meaning
6283_{16}	File cancelled	66284_{16}	Improper file format
$6A81_{16}$	Invalid function	$6A82_{16}$	File missing
$6A86_{16}$	P1 and P2 error	$6A87_{16}$	Wrong Lc

Security

ISO/IEC 7816-4 specifies an application interface for security operations for a smart card. These commands provide mechanisms through which a reader-side application can authenticate its identity to a card, a card can authenticate itself to a reader-side application, and a cardholder can authenticate his or her identity to the card. These mechanisms are used by applications to establish a known security status on a card and hence gain access to data or computational services which are protected by checks on access privileges.

Verify

Description

This command starts the comparison (in the card) of the verification data sent from the reader/terminal with the reference data stored in the card.

Command Application Protocol Data Unit

CLA	INS	Parameter 1	Parameter 2	Parameter 3
$C0_{16}$	20_{16}	00_{16}	Qualifier of reference data	Length of data field or empty

Data Field 1	Data Field 2
Verification data or empty	Empty

Response Application Protocol Data Unit

Response

2-byte status

Example of Use

APDU	Interpretation
CO_{16} 20_{16} 00_{16} 00_{16} 03_{16} 53_{16} 61_{16} 53_{16}	This command checks to see whether the card password is SAS

Error Codes

Status	Meaning	Status	Meaning
6300_{16}	Invalid verify	$63CX_{16}$	Success with retries
6983_{16}	Invalid authentication	6984_{16}	Data cancelled
$6A86_{16}$	P1 and P2 error	$6A88_{16}$	Missing data object

Internal Authenticate

Description

This command starts the computation of the authentication data by the card using the challenge data sent from the reader/terminal and a secret (key) stored in the card.

Command Application Protocol Data Unit

CLA	INS	Parameter 1	Parameter 2	Parameter 3
CO_{16}	88_{16}	ID of algorithm in card	ID of secret (key) field	Length of data

Data Field 1	Data Field 2
Challenge data	Maximum number of bytes expected in response (Le)

Response Application Protocol Data Unit

Response

Le bytes of response to the challenge plus 2-byte status

Example of Use

APDU	Interpretation
$CO_{16}88_{16}00_{16}00_{16}\ 03_{16}03_{16}02_{16}01_{16}03_{16}$	This command passes the challenge 321 from the reader to the card; it encrypts this with its known algorithm and key then returns the encrypted challenge back to the reader

Error Codes

Status	Meaning	Status	Meaning
6984_{16}	Data cancelled	6985_{16}	Invalid conditions
$6A86_{16}$	P1 and P2 error	$6A88_{16}$	Missing data object

External Authenticate

Description

This command conditionally updates the security status using the result of the computation by the card based on a challenge previously issued by the card, a (secret) key stored in the card, and authentication data supplied by the reader/terminal.

Command Application Protocol Data Unit

CLA	INS	Parameter 1	Parameter 2	Parameter 3
CO_{16}	82_{16}	ID of algorithm in card	ID of secret (key) field or empty	Length of data

Data Field 1	Data Field 2
Empty or response to challenge	Empty

Response Application Protocol Data Unit

Response
2-byte status

Example of Use

APDU	Interpretation
$CO_{16}82_{16}00_{16}00_{16}\ 03_{16}03_{16}02_{16}01_{16}03_{16}$	The card has previously generated a challenge number and sent it to the reader; the reader has encrypted it with a key it shares with the card and is now returning the encrypted challenge number to the card; if the card can validate it, then the response to this command will be a success status

Error Codes

Status	Meaning	Status	Meaning
6300_{16}	Invalid verify	$63CX_{16}$	Success with retries
6700_{16}	Incorrect Lc field	6983_{16}	Invalid authentication
6984_{16}	Data cancelled	6985_{16}	Invalid conditions
$6A86_{16}$	P1 and P2 error	$6A88_{16}$	Missing data object

Get Challenge

Description

This command forces the issuing of a challenge (such as a random number) for use in a security related procedure such as an External Authenticate command.

Command Application Protocol Data Unit

CLA	INS	Parameter 1	Parameter 2	Parameter 3
$C0_{16}$	84_{16}	0000	Empty	Le field length

Data Field 1	Data Field 2
Empty	Empty

Response Application Protocol Data Unit

Response
Le bytes of challenge data plus 2-byte status

Example of Use

APDU	Interpretation
$C0_{16}84_{16}00_{16}00_{16}06_{16}$	This command essentially asks the card to generate a 6-byte challenge string and return it to the reader-side application

Error Codes

Status	Meaning	Status	Meaning
$6A81_{16}$	Invalid function	$6A86_{16}$	P1 and P2 error

Manage Channel

Description

This command is used to open and close logical channels. A logical channel is essentially a connection between a reader-side application and a file. This is one mechanism that can be used to provide multiple reader-side applications "simultaneous" access to multiple files (applications) on the card.

Command Application Protocol Data Unit

CLA	INS	Parameter 1	Parameter 2	Parameter 3
CO_{16}	70_{16}	00_{16}=open 80_{16}=close	Channel ID 00_{16}–03_{16}	Empty or 0000_{16}

Data Field 1	Data Field 2
Empty	Empty

Response Application Protocol Data Unit

Response
Empty or logical channel number plus 2-byte status

Example of Use

APDU	Interpretation
$CO_{16}70_{16}00_{16}01_{16}$	This command causes the card to assign logical channel 1; in subsequent commands such as Select File, channel 1 can be indicated in the CLA value

Error Codes

Status	Meaning
6200_{16}	Insufficient data

Get Response

Description

This command is used to transmit from the card to the reader APDUs or parts of APDUs which otherwise would not be transmitted by the protocols in use.

Command Application Protocol Data Unit

CLA	INS	Parameter 1	Parameter 2	Parameter 3
CO_{16}	CO_{16}	00_{16}	00_{16}	The number of bytes of data to retrieve (Le)

Data Field 1	Data Field 2
Empty	Empty

Response Application Protocol Data Unit

Response

Le bytes of APDU response plus 2-byte response

Example of Use

APDU	Interpretation
CO_{16} CO_{16} 00_{16} 00_{16} 14_{16}	Retrieve the 20 bytes of information created when a `Select File` command is issued for a directory

Error Codes

Status	Meaning	Status	Meaning
$61XX_{16}$	Processing okay	6281_{16}	Return data error
6700_{16}	Incorrect Le field	$6A86_{16}$	P1 and P2 error
$6CXX_{16}$	Wrong Le length		

Envelope

Description

This command is used to transmit from the reader to the card APDUs or parts of APDUs that otherwise would not be transmitted by the protocols in use. Specifically, this allows a complete APDU to be encapsulated in the body of this APDU. This is necessary if you want to make use of secure messaging when using the T=0 link-level protocol.

Command Application Protocol Data Unit

CLA	INS	Parameter 1	Parameter 2	Parameter 3
CO_{16}	$C2_{16}$	00_{16}	00_{16}	Length of data field (Lc)

Data Field 1	Data Field 2
Encapsulated APDU	Empty or length (Le)

Response Application Protocol Data Unit

Response

Empty or part of APDU plus 2-byte status

Example of Use

APDU	Interpretation
CO_{16} $C2_{16}$ 00_{16} 00_{16} 07_{16} CO_{16} $A4_{16}$ 00_{16} 00_{16} 02_{16} $3F_{16}$ 00_{16}	This command encapsulates a Select File APDU inside it

Error Codes

Status	Meaning
6700_{16}	Incorrect Lc field

THE MULTIFLEX COMMAND SET

This appendix contains a detailed description of each of the 21 commands implemented in the Multiflex 3K operating system and to which the Multiflex 3K smart card responds. The tables here are no more a substitute for a complete documentation set for the Multiflex card than is Chapter 5, "The Schlumberger Multiflex Smart Card." They do, however, provide enough information for you to begin to experiment with each command and with the card.

Unused and RFU (reserved for future use) fields should always be filled with the default byte FF_{16} rather than the more customary default value 00_{16}. Because writing a 00_{16} to a location that contains FF_{16} requires an EEPROM erase operation, it is slower than writing an FF_{16} to a location that contains 00_{16} which only requires a write operation. Throughout the appendix, the ASCII character set is assumed.

Change PIN

Description

Replaces the 8-byte PIN in the currently selected PIN file with a new 8-byte value.

Command Application Protocol Data Unit

CLA	INS	Parameter 1	Parameter 2	Parameter 3
$F0_{16}$	24_{16}	00_{16}	01_{16}	10_{16}

Data Field 1	Data Field 2
The 8 bytes of the current value of the PIN	The 8 bytes of the new value of the PIN

Response Application Protocol Data Unit

Response

2-byte status

Example of Use

APDU	Interpretation
$F0_{16}\ 24_{16}\ 00_{16}\ 01_{16}\ 10_{16}\ 62_{16}$ $65_{16}\ 66_{16}\ 6F_{16}\ 72_{16}\ 65_{16}\ FF_{16}\ FF_{16}$ $61_{16}\ 66_{16}\ 74_{16}\ 65_{16}\ 72_{16}\ FF_{16}\ FF_{16}\ FF_{16}$	Changes the PIN in the currently selected PIN file from before to after

Status Word Return

Value	Description
6300_{16}	PIN rejected; failed attempts counter decremented
6581_{16}	Update impossible
$67XX_{16}$	Incorrect Parameter 3 value; expected value was XX_{16}
6981_{16}	No PIN defined
6983_{16}	PIN currently blocked
$6E00_{16}$	Unknown CLA
$6F00_{16}$	Internal problem with no additional information given
9000_{16}	Command executed successfully; failed attempts counter set to maximum value

Create File

Description

Creates a new file in the current directory. The new file becomes the current file.

Command Application Protocol Data Unit

CLA	INS	Parameter 1	Parameter 2	Parameter 3
$F0_{16}$	$E0_{16}$	Initialization flag 00_{16}—Initialize FF_{16}—Do not initialize	Number of records for record files; ignored otherwise	Sum of the lengths of the following two fields

Data Field 1	Data Field 2
Description of the file to be created (See Chapter 5)	The first 6 bytes of the encryption of the response to the immediately preceding Get Challenge command if the directory in which the file is being created specifies protected-mode access for this command

Response Application Protocol Data Unit

Response
2-byte status

Example of Use

APDU	Interpretation
$F0_{16}$ $E0_{16}$ FF_{16} 00_{16} 10_{16} 0000_{16} 0017_{16} 0000_{16} $F4_{16}$ FF_{16} 44_{16} 01_{16} 03_{16} $F0_{16}$ FF_{16} 00_{16}	Create a PIN file that can be updated, invalidated, and rehabilitated only by external authorization key 0 and can never be read

Status Word Return

Value	Description
6283_{16}	Current directory is invalidated
6300_{16}	Invalid protected-mode cryptogram

continues

Value	Description
6500_{16}	Too much data for protected-mode
6581_{16}	Memory problem
$67XX_{16}$	Incorrect Parameter 3 value; expected value was XX_{16}
6981_{16}	No PIN or key defined
6982_{16}	Access condition not fulfilled
6985_{16}	No Get Challenge immediately preceding command
$6A80_{16}$	File ID already in use in this directory
$6A80_{16}$	Type of current file is inconsistent with the command
$6A80_{16}$	Record length value is too large
$6A84_{16}$	Insufficient memory space available
$6B00_{16}$	Incorrect Parameter 1 or Parameter 2
$6D00_{16}$	Unknown INS
$6E00_{16}$	Unknown CLA
$6F00_{16}$	Internal problem with no additional information given
9000_{16}	Command executed successfully

Create Record

Description

Creates a new record at the end of the current record file and optionally writes data into it.

Command Application Protocol Data Unit

CLA	INS	Parameter 1	Parameter 2	Parameter 3
$C0_{16}$	$E2_{16}$	00_{16}	00_{16}	Sum of the lengths of the following two fields

Data Field 1	Data Field 2
Data to be written to new record followed by cryptogram if the directory in which the file is being created specifies protected-mode access for this command	The first 6 bytes of the encryption of the response to the immediately preceding Get Challenge command if the directory in which the file is being created specifies protected-mode authentication for the Create File command

Response Application Protocol Data Unit

Response

2-byte status

Example of Use

APDU	Interpretation
CO_{16} $E2_{16}$ 00_{16} 00_{16} 09_{16} 63_{16} 61_{16} $6D_{16}$ 62_{16} 72_{16} 69_{16} 64_{16} 67_{16} 65_{16}	Create a new record in the current record file and write `Cambridge` into it

Status Word Return

Value	Description
6283_{16}	Current file is invalidated
6300_{16}	Invalid protected-mode cryptogram
6500_{16}	Too much data for protected-mode
6581_{16}	Memory problem
$67XX_{16}$	Incorrect Parameter 3 value; expected value was XX_{16}
6981_{16}	No PIN or key defined
6982_{16}	Access condition not fulfilled
6985_{16}	No `Get Challenge` immediately preceding command
$6A80_{16}$	Type of current file is inconsistent with the command
$6A83_{16}$	Record index out of range
$6A84_{16}$	Insufficient memory space available
$6D00_{16}$	Unknown INS
$6E00_{16}$	Unknown CLA
$6F00_{16}$	Internal problem with no additional information given
9000_{16}	Command executed successfully

Decrease

Description

The oldest (that is, previous) record in a cyclic file is overwritten with the newest (that is, current) record, minus the amount given in the command. This new record then becomes the current record.

Command Application Protocol Data Unit

CLA	INS	Parameter 1	Parameter 2	Parameter 3
$F0_{16}$	30_{16}	00_{16}	00_{16}	03_{16}, the length of the following value to be subtracted, if protected-mode authentication is not required OR 09_{16}, the length of the 3-byte value plus the length of the 6-byte cryptogram, if protected-mode authentication is required

Data Field 1	Data Field 2
3-byte value to be subtracted from the current record	The first 6 bytes of the encryption of the response to the immediately preceding Get Challenge command if the directory in which the file is being created specifies protected-mode authentication for the Create File command

Response Application Protocol Data Unit

Response

2-byte status

Example of Use

APDU	Interpretation
$F0_{16}$ 30_{16} 00_{16} 00_{16} 03_{16} 00_{16} 00_{16} 01_{16}	Subtract 1 from the current record in a cyclic file and overwrite the oldest record in the file with this new value

Status Word Return

Value	Description
$61XX_{16}$	Command executed successfully; XX_{16} bytes of response data are available
6283_{16}	Currently selected file is invalidated
6300_{16}	Invalid protected-mode cryptogram
6500_{16}	Too much data for protected-mode
6581_{16}	Update impossible
$67XX_{16}$	Incorrect Parameter 3 value; expected value was XX_{16}

Value	Description
6981_{16}	No PIN and or key defined
6986_{16}	Currently selected file is not a cyclic file
$6A80_{16}$	Type of current file is inconsistent with the command
$6D00_{16}$	Unknown INS
$6E00_{16}$	Unknown CLA
$6F00_{16}$	Internal problem with no additional information given
9850_{16}	Decrease cannot be performed; new value would be less than minimum value

Delete File

Description

Deletes the named file.

Command Application Protocol Data Unit

CLA	INS	Parameter 1	Parameter 2	Parameter 3
$F0_{16}$	$E4_{16}$	00_{16}	00_{16}	02_{16}, the length of the following file ID, if protected mode authentication is not required; or 08_{16}, the length of the file ID plus the length of the 6-byte cryptogram if protected-mode authentication is required

Data Field 1	Data Field 2
2-byte file identifier	The first 6 bytes of the encryption of the response to the immediately preceding Get Challenge command if the currently selected file specifies protected-mode authentication for this command

Response Application Protocol Data Unit

Response
2-byte status

Example of Use

APDU	Interpretation
$F0_{16}$ $E4_{16}$ 00_{16} 00_{16} 02_{16} 00_{16} 33_{16}	Delete the file with the file ID 0033_{16} in the currently selected directory

Status Word Return

Value	Description
6283_{16}	Current file is invalidated
6300_{16}	Invalid protected-mode cryptogram
6500_{16}	Too much data for protected-mode
6581_{16}	Memory problem
$67XX_{16}$	Incorrect Parameter 3 value; expected value was XX_{16}
6981_{16}	No PIN or key defined
6982_{16}	Access condition not fulfilled
6985_{16}	No Get Challenge immediately preceding command
$6A80_{16}$	Type of current file is inconsistent with the command
$6A82_{16}$	File ID not found
$6B00_{16}$	Incorrect Parameter 1 or Parameter 2
$6D00_{16}$	Unknown INS
$6E00_{16}$	Unknown CLA
$6F00_{16}$	Internal problem with no additional information given
9000_{16}	Command executed successfully

External Authentication

Description

The terminal wishes to gain external authentication access to the card without sending a key to it using Verify Key. It got a challenge from the card using Get Challenge and is now going to return its encryption of this challenge to prove it knows the key.

Command Application Protocol Data Unit

CLA	INS	Parameter 1	Parameter 2	Parameter 3
$C0_{16}$	82_{16}	00_{16}	00_{16}	07_{16}

Data Field 1	Data Field 2
Key Number: 00_{16}, 01_{16}, 02_{16}, ..., 09_{16}, $0A_{16}$..., $0E_{16}$, $0F_{16}$	First 6 bytes of the encryption of the the challenge provided by the card as a response to the immediately preceding Get Challenge command

Response Application Protocol Data Unit

Response

2-byte status

Example of Use

APDU	Interpretation
CO_{16} 82_{16} 00_{16} 00_{16} 07_{16} 03_{16} $3E_{16}$ 67_{16} $A8_{16}$ 45_{16} 91_{16} $7C_{16}$	The first 6 bytes of the encryption of the challenge provided by the card using key 3 in the external authentication file associated with the current directory are $3E_{16}$ 67_{16} $A8_{16}$ 45_{16} 91_{16} $7C_{16}$

Status Word Return

Value	Description
6300_{16}	External authentication failed; failed attempts counter decremented
$67XX_{16}$	Incorrect Parameter 3 value; expected value was XX_{16}
6981_{16}	No key defined
6983_{16}	Key blocked
6985_{16}	No Get Challenge immediately preceding command
$6D00_{16}$	Unknown INS
$6E00_{16}$	Unknown CLA
$6F00_{16}$	Internal problem with no additional information given
9000_{16}	Command executed successfully; failed attempts counter set to maximum value

Get Challenge

Description

The card is requested to send back an 8-byte challenge.

Command Application Protocol Data Unit

CLA	INS	Parameter 1	Parameter 2	Parameter 3
CO_{16}	84_{16}	00_{16}	00_{16}	08_{16}

Data Field 1	Data Field 2
N/A	N/A

Response Application Protocol Data Unit

Response

8 bytes of challenge if the command is successful followed by the 2-byte status

Example of Use

APDU	Interpretation
CO_{16} 84_{16} 00_{16} 00_{16} 08_{16}	The card returns a 10-byte response consisting of an 8-byte challenge followed by the normal 2-byte status code

Status Word Return

Value	Description
$67XX_{16}$	Incorrect Parameter 3 value; expected value was XX_{16}
$6D00_{16}$	Unknown INS given in the command
$6E00_{16}$	Unknown CLA given in the command
$6F00_{16}$	Internal problem with no additional information given
9000_{16}	Command executed successfully

Get Response

Description

Retrieves data typically created by the immediately preceding command from the card.

Command Application Protocol Data Unit

CLA	INS	Parameter 1	Parameter 2	Parameter 3
CO_{16}	CO_{16}	00_{16}	00_{16}	The number of bytes of data to retrieve

Data Field 1	Data Field 2
N/A	N/A

Response Application Protocol Data Unit

Response

The number of bytes requested followed by the 2-byte status

Example of Use

APDU	Interpretation
$CO_{16}\ CO_{16}\ 00_{16}\ 00_{16}\ 14_{16}$	Retrieve the 20 bytes of information created when a `Select File` command is issued for a directory.

Status Word Return

Value	Description
$67XX_{16}$	Incorrect Parameter 3 value; expected value was XX_{16}
$6D00_{16}$	Unknown INS
$6E00_{16}$	Unknown CLA
$6F00_{16}$	Internal problem with no additional information given
9000_{16}	Command executed successfully

Increase

Description

The oldest (i.e., previous) record in a cyclic file is overwritten with the newest (i.e., current) record, plus the amount given in the command. This new record then becomes the current record.

Command Application Protocol Data Unit

CLA	INS	Parameter 1	Parameter 2	Parameter 3
FO_{16}	32_{16}	00_{16}	00_{16}	03_{16}, the length of the following value to be added, if protected-mode authentication is not required; or 09_{16}, the length of the value plus the length of the cryptogram if protected-mode authentication is required

Data Field 1	Data Field 2
3-byte numeric value to be added on to the current record	The first 6 bytes of the encryption of the response to the immediately preceding `Get Challenge` command if the directory in which the file is being created specifies protected-mode authentication for the `Create File` command

Response Application Protocol Data Unit

Response

2-byte status

Example of Use

APDU	Interpretation
FO_{16} 32_{16} 00_{16} 00_{16} 03_{16} 00_{16} 00_{16} 02_{16}	Add 2 to the value in the last 6 bytes of the current record in the currently selected cyclic file and write the record thereby created over the record previous to the current one in the cyclic file; make this overwritten record the current record

Status Word Return

Value	Description
$61XX_{16}$	Command executed successfully; XX_{16} bytes of response data are available
6283_{16}	Currently selected file is invalidated
6300_{16}	Invalid protected-mode cryptogram
6500_{16}	Too much data for protected-mode
6581_{16}	Update impossible
$67XX_{16}$	Incorrect Parameter 3 value; expected value was XX_{16}
6981_{16}	No PIN or key defined
6986_{16}	Currently selected file is not an elementary file
$6A80_{16}$	Type of currently selected file is inconsistent with the instruction
$6D00_{16}$	Unknown INS
$6E00_{16}$	Unknown CLA
$6F00_{16}$	Internal problem with no additional information given
9850_{16}	Increase cannot be performed; new value would be greater than maximum value

Internal Authentication

Description

The terminal wishes to authenticate the card to ensure it is a valid card, so it sends the card a challenge that the card must encrypt using a specified key in the internal authorization file (0001_{16}) for the current directory. A following Get Response command returns the first 6 bytes of the card's encryption of the challenge using the indicated key.

Command Application Protocol Data Unit

CLA	INS	Parameter 1	Parameter 2	Parameter 3
$C0_{16}$	88_{16}	00_{16}	Key number: 00_{16}, 01_{16}, 02_{16}, ..., 09_{16}, $0A_{16}$, ..., $0E_{16}$, $0F_{16}$	08_{16}

Data Field 1	Data Field 2
8-byte challenge	N/A

Response Application Protocol Data Unit

Response
2-byte status

Example of Use

APDU	Interpretation
$C0_{16}$ 88_{16} 00_{16} 03_{16} 08_{16} 64_{16} 69_{16} 73_{16} $6B_{16}$ 65_{16} 74_{16} 74_{16} 65_{16}	The terminal sends the challenge diskette to the card and expects it to be encrypted with key 3 in the internal authorization key file (0001_{16}) associated with the current directory

Status Word Return

Value	Description
$61XX_{16}$	Command executed successfully; XX_{16} bytes of response data are available
$67XX_{16}$	Incorrect Parameter 3 value; expected value was XX_{16}
6981_{16}	No PIN or key defined
6982_{16}	Access condition not fulfilled
$6B00_{16}$	Incorrect Parameter 1 or Parameter 2
$6D00_{16}$	Unknown INS
$6E00_{16}$	Unknown CLA
$6F00_{16}$	Internal problem with no additional information given

Invalidate

Description

The currently selected elementary file is invalidated and will subsequently only respond successfully to the Select File and Rehabilitate commands.

Command Application Protocol Data Unit

CLA	INS	Parameter 1	Parameter 2	Parameter 3
$F0_{16}$	04_{16}	00_{16}	00_{16}	00_{16} if issuing the Invalidate command for this file does not require protected-mode access; or 06_{16}, the length of the following cryptogram, if it does

Data Field 1	Data Field 2
The first 6 bytes of the encryption of response to immediately preceding Get Challenge command if the currently selected file requires protected-mode authentication for this command	N/A

Response Application Protocol Data Unit

Response
2-byte status

Example of Use

APDU	Interpretation
$F0_{16}\ 04_{16}\ 00_{16}\ 00_{16}\ 00_{16}$	The currently selected file which does not require protected-mode authentication to be invalidated is invalidated

Status Word Return

Value	Description
6283_{16}	Current file is already invalidated
6300_{16}	Invalid protected-mode cryptogram

Value	Description
6581_{16}	Memory problem
$67XX_{16}$	Incorrect Parameter 3 value; expected value was XX_{16}
6981_{16}	No PIN or key defined
6982_{16}	Access condition not fulfilled
6985_{16}	No Get Challenge immediately preceding command
6986_{16}	No file selected
$6D00_{16}$	Unknown INS
$6E00_{16}$	Unknown CLA
$6F00_{16}$	Internal problem with no additional information given
9000_{16}	Command executed successfully

Read Binary

Description

Reads a sequence of bytes from the currently selected transparent file.

Command Application Protocol Data Unit

CLA	INS	Parameter 1	Parameter 2	Parameter 3
$C0_{16}$	$B0_{16}$	High byte of the 2-byte offset number	Low byte of the 2-byte offset number	Number of bytes to read starting at the offset byte

Data Field 1	Data Field 2
N/A	N/A

Response Application Protocol Data Unit

Response

The number of bytes requested followed by the 2-byte status

Example of Use

APDU	Interpretation
$C0_{16}\ B0_{16}\ 00_{16}\ 00_{16}\ 10_{16}$	Read 16 bytes from the currently selected transparent file starting with the first byte in the file

Status Word Return

Value	Description
6283_{16}	Currently selected file is invalidated
6300_{16}	Invalid protected-mode cryptogram
6581_{16}	Memory problem
$67XX_{16}$	Incorrect Parameter 3 value; expected value was XX_{16}
6981_{16}	No PIN or key defined
6986_{16}	No currently selected elementary file
$6A80_{16}$	Current file type is inconsistent with the instruction
$6B00_{16}$	Offset out of range
$6D00_{16}$	Unknown INS
$6E00_{16}$	Unknown CLA
$6F00_{16}$	Internal problem with no additional information given
9000_{16}	Command executed successfully

Read Record

Description

Reads one record from the currently selected record file.

Command Application Protocol Data Unit

CLA	INS	Parameter 1	Parameter 2	Parameter 3
$C0_{16}$	$B2_{16}$	Index of the record to be read (01_{16}, 02_{16}, ..., FF_{16}) OR 00_{16} if the current record is to be read	Selection of record to be read: 00 first record; 01 last record; 02 next record; 03 previous record; 04 current record; if index is 0 or index record, if it isn't	The number of bytes to be read from the record identified by Parameter 1 and Parameter 2; must be equal to the length of the record in the file

Data Field 1	Data Field 2
N/A	N/A

Response Application Protocol Data Unit

Response

The number of bytes in the record if the command is successful followed by the usual byte status

Example of Use

APDU	Interpretation
CO_{16} $B2_{16}$ 06_{16} 04_{16} 14_{16}	The records in the selected fixed-length record file are 20 bytes long; this command reads the sixth record in the file

Status Word Return

Value	Description
6281_{16}	Data may be corrupted
6283_{16}	Currently selected file is invalidated
6581_{16}	Memory problem
$67XX_{16}$	Incorrect Parameter 3 value; expected value was XX_{16}
6981_{16}	No PIN or key defined
6986_{16}	Currently selected file is not an elementary file
$6A80_{16}$	Current file type is inconsistent with the instruction
$6A83_{16}$	Out of range/record not found
$6B00_{16}$	Incorrect Parameter 1 or Parameter 2
$6D00_{16}$	Unknown INS
$6E00_{16}$	Unknown CLA
$6F00_{16}$	Internal problem with no additional information given
9000_{16}	Command executed successfully

Rehabilitate

Description

The currently selected elementary file is rehabilitated (that is, removed from invalidated status).

Command Application Protocol Data Unit

CLA	INS	Parameter 1	Parameter 2	Parameter 3
$F0_{16}$	44_{16}	00_{16}	00_{16}	00_{16} if protected-mode authentication is not required;
				or
				06_{16}, the length of the following 6-byte cryptogram, if protected-mode authentication is required

Data Field 1	Data Field 2
The first 6 bytes of the encryption of the response to the immediately preceding Get Challenge command if the currently selected file specifies protected-mode authentication	N/A

Response Application Protocol Data Unit

Response

2-byte status

Example of Use

APDU	Interpretation
$F0_{16}$ 44_{16} 00_{16} 00_{16} 06_{16} 34_{16} $8D_{16}$ $C1_{16}$ 22_{16} $A7_{16}$ 58_{16}	Rehabilitate the currently selected file where 33_{16} $8D_{16}$ $C1_{16}$ 22_{16} $A7_{16}$ and 58_{16} are the first 6 bytes of the encryption of the challenge that was just previously retrieved from the card using Get Challenge

Status Word Return

Value	Description
6283_{16}	File is not invalidated
6300_{16}	Invalid protected-mode cryptogram
6500_{16}	Too much data for protected-mode
6581_{16}	Memory problem

Value	Description
$67XX_{16}$	Incorrect Parameter 3 value; expected value was XX_{16}
6981_{16}	No PIN or key defined
6982_{16}	Access condition not fulfilled
6985_{16}	No Get Challenge immediately preceding command
6986_{16}	No file selected
$6B00_{16}$	Incorrect Parameter 1 or Parameter 2
$6D00_{16}$	Unknown INS
$6E00_{16}$	Unknown CLA
$6F00_{16}$	Internal problem with no additional information given
9000_{16}	Command executed successfully

Seek

Description

Locate a record in a linear record file by matching a pattern of characters to the characters in each record starting at a given offset from the beginning of the record.

Command Application Protocol Data Unit

CLA	INS	Parameter 1	Parameter 2	Parameter 3
$F0_{16}$	$A2_{16}$	$Offset_{16}$	Search mode 00_{16} from first record; 02_{16} from next record	Number of characters in the following pattern

Data Field 1	Data Field 2
Character string to be matched	N/A

Response Application Protocol Data Unit

Response

2-byte status

Example of Use

APDU	Interpretation
$F0_{16}\ A2_{16}\ 08_{16}\ 02_{16}\ 04_{16}\ 62_{16}$ $6F_{16}\ 6F_{16}\ 6B_{16}$	Continue searching from the record after the current one for the text string 'book', starting at the ninth character in each record

Status Word Return

Value	Description
6281_{16}	Data may be corrupted
6283_{16}	The file at the current pointer is invalidated
$67XX_{16}$	Incorrect Parameter 3 value; expected value was XX_{16}
6981_{16}	No PIN or key defined
6982_{16}	Access condition not fulfilled
6986_{16}	Currently selected file is not a linear record file
$6A80_{16}$	Pattern not found
$6B00_{16}$	Offset out of range
$6D00_{16}$	Unknown INS
$6E00_{16}$	Unknown CLA
$6F00_{16}$	Internal problem with no additional information given

Select File

Description

The file whose file ID is given in the data field of the command becomes the currently selected file. It must be a file in the currently selected directory. If the named file is a directory, then it becomes the currently selected directory.

Command Application Protocol Data Unit

CLA	INS	Parameter 1	Parameter 2	Parameter 3
$C0_{16}$	$A4_{16}$	00_{16}	00_{16}	02_{16}

Data Field 1	Data Field 2
2-byte file identifier	N/A

Response Application Protocol Data Unit

Response

2-byte status; if the high-order byte of the status word is 61_{16} then the low-order byte is the number of bytes of file description data that can be retrieved with a subsequent Get Response command

Example of Use

APDU	Interpretation
CO_{16} $A4_{16}$ 00_{16} 00_{16} 02_{16} $3F_{16}$ 00_{16}	The master file becomes the currently selected directory

Status Word Return

Value	Description
$61XX_{16}$	Command executed successfully; XX_{16} bytes of response data are available
6281_{16}	Data may be corrupted
$67XX_{16}$	Incorrect Parameter 3 value; expected value was XX_{16}
$6A82_{16}$	File with given file ID not found in current directory
$6D00_{16}$	Unknown INS
$6E00_{16}$	Unknown CLA
$6F00_{16}$	Internal problem with no additional information given

Unblock PIN

Description

The selected PIN file has become blocked because the number of presentations of an incorrect PIN has exceeded the number of allowed tries. This command will unblock the PIN file and reset the PIN to a new value.

Command Application Protocol Data Unit

CLA	INS	Parameter 1	Parameter 2	Parameter 3
$F0_{16}$	$2C_{16}$	00_{16}	01_{16}	10_{16}

Data Field 1	Data Field 2
8-byte unblocking PIN for current PIN file	8-byte new PIN

Response Application Protocol Data Unit

Response
2-byte status

Example of Use

APDU	Interpretation
$F0_{16}$ $2C_{16}$ 00_{16} 01_{16} 10_{16} 38_{16} 37_{16} 36_{16} 35_{16} 34_{16} 33_{16} 32_{16} 31_{16} 35_{16} 36_{16} 37_{16} 38_{16} FF_{16} FF_{16} FF_{16} FF_{16}	The unblocking key is 87654321; the PIN file is unblocked and the new PIN set to 5678

Status Word Return

Value	Description
6300_{16}	Unblocking key rejected; failed attempts counter decremented
6581_{16}	Update impossible
$67XX_{16}$	Incorrect Parameter 3 value; expected value was XX_{16}
6981_{16}	No PIN defined
6983_{16}	The unblocking key is blocked
$6D00_{16}$	Unknown INS
$6E00_{16}$	Unknown CLA
$6F00_{16}$	Internal problem with no additional information given
9000_{16}	Command executed successfully; failed attempts counter set to maximum value

Update Binary

Description

A sequence of bytes is written into the currently selected transparent elementary file.

Command Application Protocol Data Unit

CLA	INS	Parameter 1	Parameter 2	Parameter 3
$C0_{16}$	$D6_{16}$	High byte of the 2-byte offset number	Low byte of the 2-byte offset number	The number of bytes to be written into the file starting at the offset byte plus 6 if a protected-mode cryptogram is provided

Data Field 1	Data Field 2
The data bytes to be written into the transparent file starting at the offset byte	The first 6 bytes of the encryption of the response to the immediately preceding Get Challenge command if the currently selected file specifies protected-mode authentication for this command

Response Application Protocol Data Unit

Response

2-byte status

Example of Use

APDU	Interpretation
$C0_{16}$ $D6_{16}$ 00_{16} 00_{16} 17_{16} FF_{16} FF_{16} FF_{16} 31_{16} 32_{16} 33_{16} 34_{16} FF_{16} FF_{16} FF_{16} FF_{16} 03_{16} 03_{16} 38_{16} 37_{16} 36_{16} 35_{16} 34_{16} 33_{16} 32_{16} 31_{16} 03_{16} 03_{16}	Suppose the currently selected transparent file is the PIN file (0001) in the currently selected directory; this Update Binary command sets the PIN to 1234 and the Unblock PIN to 87654321 both with tries set to 3

Status Word Return

Value	Description
6283_{16}	The file at the current pointer is invalidated
6300_{16}	Invalid protected-mode cryptogram
6500_{16}	Too much data for protected-mode
6581_{16}	Update impossible
$67XX_{16}$	Incorrect Parameter 3 value; expected value was XX_{16}
6981_{16}	No PIN or key defined
6982_{16}	Access condition not fulfilled
6985_{16}	No Get Challenge immediately preceding command
6986_{16}	No currently selected file
$6A80_{16}$	Current file type is inconsistent with the instruction
$6B00_{16}$	Offset out of range

continues

Value	Description
$6D00_{16}$	Unknown INS
$6E00_{16}$	Unknown CLA
$6F00_{16}$	Internal problem with no additional information given
9000_{16}	Command executed successfully

Update Record

Description

One record in the currently selected record file is overwritten with new data.

Command Application Protocol Data Unit

CLA	INS	Parameter 1	Parameter 2	Parameter 3
CO_{16}	DC_{16}	Index of the record to be overwritten $(01_{16}, 02_{16}, ..., FF_{16})$ OR 0 if the current record is to be overwritten	Selection of record to be overwritten: 00 first record; 01 last record; 02 next record; 03 previous record; 04 current record; if index is 0 or index record if it isn't	The number of bytes to be written into the record identified by Parameter 1 and Parameter 2; must be equal to the length of the record in the file; add 6 if a protected-mode cryptogram is provided

Data Field 1	Data Field 2
The data bytes to be written into the record identified by Parameter 1 and Parameter 2	The first 6 bytes of the encryption of the response to the immediately preceding Get Challenge command if the currently selected file specifies protected-mode authentication for this command

Response Application Protocol Data Unit
Response

2-byte status

Example of Use

APDU	Interpretation
$C0_{16}$ DC_{16} 06_{16} 04_{16} 14_{16} 53_{16} 61_{16} $6C_{16}$ $6C_{16}$ 79_{16} 20_{16} 47_{16} 72_{16} 65_{16} 65_{16} $6E_{16}$ 00_{16} 00_{16} 00_{16} 00_{16} 00_{16} 00_{16} 00_{16} 00_{16} 00_{16}	The records in the selected fixed-length record file are 20 bytes long; this command writes Sally Green into the sixth record in this file

Status Word Return

Value	Description
6283_{16}	Currently selected file is invalidated
6300_{16}	Invalid protected-mode cryptogram
6500_{16}	Too much data for protected-mode
6581_{16}	Update impossible
$67XX_{16}$	Incorrect Parameter 3 value; expected value was XX_{16}
6981_{16}	No PIN or key defined
6982_{16}	Access condition not fulfilled
6985_{16}	No Get Challenge immediately preceding command
6986_{16}	No EF selected as current
$6A80_{16}$	Current file type is inconsistent with the instruction
$6A83_{16}$	Out of range/record not found
$6B00_{16}$	Incorrect Parameter 1 or Parameter 2
$6D00_{16}$	Unknown INS
$6E00_{16}$	Unknown CLA
$6F00_{16}$	Internal problem with no additional information given
9000_{16}	Command executed successfully

Verify PIN

Description

Attempt to match the 8 bytes in the command with the 8-byte PIN in the PIN file for the current directory. If the match is exact, then PIN access privileges are granted.

Command Application Protocol Data Unit

CLA	INS	Parameter 1	Parameter 2	Parameter 3
CO_{16}	20_{16}	00_{16}	01_{16}	08_{16}

Data Field 1	Data Field 2
8-byte PIN to be matched against the 8-byte PIN on the card	N/A

Response Application Protocol Data Unit

Response

2-byte status

Example of Use

APDU	Interpretation
$CO_{16}\ 20_{16}\ 00_{16}\ 01_{16}\ 08_{16}\ 31_{16}$ $32_{16}\ 33_{16}\ 34_{16}\ FF_{16}\ FF_{16}\ FF_{16}\ FF_{16}$	Presentation of the PIN code 1234

Status Word Return

Value	Description
6300_{16}	PIN authentication failed
$67XX_{16}$	Incorrect Parameter 3 value; expected value was XX_{16}
6981_{16}	No PIN defined
6983_{16}	PIN currently blocked
$6D00_{16}$	Unknown INS
$6E00_{16}$	Unknown CLA
$6F00_{16}$	Internal problem with no additional information given
9000_{16}	Command executed successfully

Verify Key

Description

Match a byte sequence with a key in the external authorization file (0011_{16}) for the current directory. If the match is exact, external authorization access privileges are granted.

Command Application Protocol Data Unit

CLA	INS	Parameter 1	Parameter 2	Parameter 3
$F0_{16}$	$2A_{16}$	00_{16}	Key Number: 00_{16}, 01_{16}, 02_{16}, ..., 09_{16}, $0A_{16}$, ..., $0E_{16}$, $0F_{16}$	Length of the following key

Data Field 1	Data Field 2
Key to be presented to the external authorization file (0011_{16}) for the current directory	N/A

Response Application Protocol Data Unit

Response
2-byte status

Example of Use

APDU	Interpretation
$F0_{16}$ $2A_{16}$ 00_{16} 01_{16} 47_{16} 46_{16} 58_{16} 49_{16} 32_{16} 56_{16} 78_{16} 40_{16}	Match the transportation key in the Multiflex card included in this book

Status Word Return

Value	Description
6300_{16}	Key verify rejected
$67XX_{16}$	Incorrect Parameter 3 value; expected value was XX_{16}
6981_{16}	No key defined
6983_{16}	Key blocked
$6A82_{16}$	File not found
$6B00_{16}$	Incorrect Parameter 1 or Parameter 2
$6D00_{16}$	Unknown INS
$6E00_{16}$	Unknown CLA
$6F00_{16}$	Internal problem with no additional information given
9000_{16}	Command executed successfully

Glossary

A3 and A8 Two cryptographic algorithms used in GSM cellular telephony and typically implemented in GSM SIM smart cards.

ABS (acrylonitrile butadiene styrene) A common plastic material used for the manufacture of smart cards.

AC (access condition) An attribute in a file header that allows or denies execution of certain commands based on certain security conditions such as authentication of the entity attempting to execute the command.

AID (application identifier) A unique number assigned to smart card applications.

algorithm A set of detailed instructions for performing a mathematical operation.

alt.technology.smartcards A Usenet newsgroup devoted to smart cards.

ANSI (American National Standards Institute) An American technical standards body and the representative of the United States to the International Standards Organization (ISO).

anticollision When using a contactless smart card, the data being transmitted from the card to the reader doesn't collide or interfere with the data being transmitted from the reader to the card.

APDU (application protocol data unit) A unit of data transfer between a smart card and an application program; a smart card command or command response.

API (application programming interface) Contains calls a program can make on routines stored in a function library or implemented in the operating system.

Arimura, Dr. Kunitaka The Japanese inventor who received a patent on smart cards in 1971.

ASC (application-specific command) An extension of the basic smart card operating system, often stored in the smart card EEPROM.

ASCII (American Standard Code for Information Interchange) A method of digitally representing characters in the Latin alphabet using 1 byte or 8 bits. For example, 6116 is the ASCII representation of lowercase Latin letter *a*. *See also* Unicode.

asynchronous protocol A mode of data transmission in which the transmission start time of a character or block of characters is arbitrary. *See also* synchronous protocol.

ATR (answer to reset) A data string returned by a smart card when the microprocessor in the card is physically reset. Two types of data strings are standardized: They are described as asynchronous transfer protocols T=0 and T=1.

authenticate To establish the identity of the origination or originator of a transaction or other data-processing request.

authorize To grant privileges typically to access data, usually based on successful authentication.

batch card A smart card that carries a key that enables its holder to unlock a shipment or batch of other smart cards. A batch card carries a transport key. *See also* mother card.

biometrics The use of a person's physical characteristics such as fingerprints, hand geometry, voice or signature characteristics, eye patterns, and so on, for authentication.

black book A catalog of information used to subvert smart card security systems.

blinding Taking provisions in a smart card's operation to defeat voltage and timing attacks. Blinding, for example, would ensure that all multiplications take the same amount of time independent of the values of the multiplier and the multiplicand.

Bright, Roy The French publicist who coined the term smart card.

byte string A sequence of bytes.

C-SET (Chip-Secured Electronic Transaction) The French version of SET, which incorporates a smart card in its specification.

CA (certification authority) An organization or enterprise that issues digital certificates, primarily those attesting to an individual's identity.

CAD (card accepting device) A smart card reader.

capture To not return a card to the cardholder if an anomalous condition is encountered before a transaction is complete. A capture reader takes the smart card completely inside its physical security perimeter so that it cannot be extracted by the user before the transaction is completed.

Card Europe A smart card industry association. See www.gold.net/users/ct96.

cardholder The person carrying and using a smart card. A cardholder does not necessarily own the card or have any rights other than holding and using it.

CARDIS An international smart card conference, Smart Card Research and Advanced Applications, held roughly every 18 months, that features academic papers on smart card research.

CardTech/Securetech A North American smart card convention held twice a year.

Carte Bancaire The smart card issued by Groupement des Cartes Bancaires, a French bankcard association.

Cartes An annual smart card convention held in Paris.

Castrucci, Paul The American inventor who received U.S. Patent 3,702,464 on a smart card in 1972.

CEN (Comité Européen de Normalisation) A European standards organization, located in Brussels.

cert Conversational shorthand for *digital certificate.*

challenge A random string of bytes sent from a data processing system to another system that it is trying to authenticate. The receiving system must encrypt the challenge with an encryption key in its possession and return the encrypted challenge to the sending system. If the sending system can decrypt the encrypted challenge, it knows the receiving system possesses the key that encrypted it and this authenticates the system to which the challenge was sent.

checksum A single numeric value computed from a large body of text or data that can be quickly recomputed by the recipient of the text and data to check if any characters in the body have been changed during transmission. Unlike a hash value, similar bodies of text may yield equal checksums. Checksums guard against random transmission errors (not deliberate attempts) to alter the content of a message.

CHV (cardholder verification) A secret number or password, known only to the cardholder, which is required to access certain services on a smart card. Also known as a personal identification number (PIN).

CLA The first data field in an ISO 7816-4 command that gives the class of the command.

CLK The contact or pad on a smart card module through which clock signals are provided to run the smart card processor.

clock rate The rate at which the clock signal provided to a smart card processor changes; typically, 5 MHz or 5,000,000 pulses per second. Smart card processors divide this by 2 and take on the average of 4 or 5 "clocks" per instruction and so run at about $\frac{1}{2}$ MIP or 500,000 instructions per second.

COMP128 An authentication algorithm popular in telecommunications and often found on GSM SIM cards.

contact card A smart card that is activated by being inserted into a smart card reader, which presses contacts against the contact pads of the smart card module. *See also* contactless card.

contactless card A smart card that is activated by being held near the smart card reader rather than being put into the reader, as with contact cards. Power is provided to the card through inductance coils and communication occurs via radio frequency signals and a capacitive plate antenna. *See also* contact card.

core The instruction set used by a smart card; for example, an 8051 core implements the Intel 8051 instruction set. It is called the core because the integrated circuit that implements the instructions is the core of the smart card integrated circuit.

COS (card operating system) The program contained in the smart card ROM that is used for communicating with the smart card, managing security, and managing data in the smart card file system.

CPU (central processing unit) The integrated circuitry on a smart card that executes the program stored on the card.

CRT (Chinese remainder theorem) A theorem about the unique factorization of integers that is used in some cryptographic algorithms.

cryptogram An encrypted block of text or random data; used in smart card security protocols to demonstrate possession of a secret key without revealing the key itself.

cryptographic coprocessor Special integrated circuits for quickly doing calculations, particularly modular arithmetic and large integer calculations, associated with cryptographic operations and algorithms. These circuits are added to a standard processor core and therefore are called coprocessors.

cyclic file A type of file on a smart card that contains records such that the first record is returned when a read next command is issued on the last record; thus, the records form a ring and cycle from one to the next.

Danmont A smart card operating system developed in Denmark and used in the VisaCash card. See www.iccard.dk.

daughter card One of a batch or shipment of cards that is unlocked with a mother card.

DEA (data encryption algorithm) Synonym for DES.

DES (data encryption standard) A secret key cryptographic algorithm defined and promoted by the U.S. government.

Dethloff, Jürgen The German co-inventor of the smart card in 1968. *See also* Gröttrupp, Helmut.

DF (dedicated file) A smart card directory file that holds other files.

digital certificate A digital message that contains the public key of an individual together with a guarantee from a certificate authority that the public key belongs to the individual.

digital signature A digital technique that authenticates the user's transaction. A digital signature can, for example, be the encryption of a hash of the transaction with the individual's private key.

diversified key A smart card key that is computed from a smart card's serial number and a master key. Diversified key techniques let every card in a large set of cards be accessed with a unique key without the necessity of maintaining a record of which key is on which card. Both the master key and the calculation program are kept in a highly secure environment.

DSA (digital signature algorithm) A cryptographic algorithm approved by the U.S. government for use in creating digital signatures.

DSS (digital signature standard) The U.S. standard that defines DSA and its use.

E-cash card A stored-value smart card that contains money in digital form in one or more national currencies such as kroner, francs, yen, marks, or dollars. When you spend money from the card, the host application decrements a currency value and when you add more money to the card, the host application increments a currency value. Don't try this at home.

EEPROM (electrically erasable programmable read-only memory) Memory in a smart card that holds its contents when power is removed, that is, when the card is removed from the card reader. Unlike with ROM, new values can be written to EEPROM by the smart card CPU. EEPROM is used to store smart card values that are set during personalization, such as account numbers or values that can change, such as the amount of value stored on the card.

EF (elementary file) An elementary file is part of the smart card file system that contains application data. *See also* DF (dedicated file), MF (master file).

EFT (electronic funds transfer) A funds transfer that is sent electronically, either by telecommunication or written on magnetic media such as tape, cassette, or disk.

electronic wallet Similar to an e-purse, with added functions such as credit and debit account access capability. *See also* EP or E-purse.

emulator A computer program plus special hardware that enables a program developer to run a smart card program on the actual smart card chip but still be able to control and analyze the execution of the program. An emulator, for example, typically allows the developer to single-step the smart card processor and examine the smart card processor's registers and memory.

EMV (Europay, MasterCard, and Visa) An alliance of bankcard associations that generated a smart card standard for payment (credit and debit) smart cards, popularly called EMV'96.

EN 726 A standard for smart cards and terminals for telecommunication use. The standard is the technical basis for smart cards in Europe.

EN 742 A standard for the contacts for cards and devices used in Europe. New editions specify the format used for the GSM subscriber identity module (SIM).

EP or E-purse (electronic purse) A smart card that stores small amounts of currency, usually less than $1,000. Some electronic purses can be reloaded; some cannot and are discarded when empty.

ESCAT (European Smart Card Application and Technology) A smart card convention held annually at the beginning of September.

ETSI (European Telecommunication Standards Institute) A European standards body.

FIPS 140-1 A U.S. federal standard titled "Security Requirements for Cryptographic Modules" that concerns physical security of smart cards when used as cryptographic devices. For more information, go to `http://csrc.ncsl.nist.gov/fips/fips140-1.txt`.

FLASH A type of nonvolatile memory that can be written much faster than EEPROM memory. Although usually written in all capital letters, FLASH is not an acronym, but rather refers to fact that the memory can be bulk erased (that is, electronically "flashed" as PROM memory of yore was flashed with UV light).

FRAM (ferroelectric memory) A type of nonvolatile memory based on electric field orientation with nearly an infinite write capability as opposed to normal EEPROM memory, which can only be written approximately 10,000 times.

FSCUG (federal smart card users group) A U.S. government smart card users group which promulgates standards and specifications for the use of smart cards in government data-processing functions.

GND The ground contact or pad on a smart card module.

Gröttrupp, Helmut The German co-inventor of the smart card in 1968. *See also* Dethloff, Jürgen.

GSCAS (Global Smart Card Advisory Service) A smart card consulting service. For more information, go to `www.gscas.com`.

GSM (Groupe Spécial Mobile or Global Service for Mobile Communications) A European cellular telephone standard. GSM telephones use smart cards called SIM cards to store subscriber account information.

handshake A protocol between two devices, such as a smart card and a personal computer, to establish a common dialog.

hard mask *See* mask.

hash A string of bytes of a fixed length that is effectively a unique representation of a longer document. *Effectively unique* means that it is difficult to find another document that produces the same hash value and that any slight change in the long document will produce a different hash value.

hybrid card A smart card that can function as more than one kind of card—for example, a smart card that can function as both a contact and a contactless card. Or, a smart card that also has a magnetic stripe or a barcode.

I/O (input/output) The input/ouput contact or pad on a smart card module though which messages are passed to and received from the microprocessor in the card.

IC (integrated circuit) A small electronic device made from metallic and semiconductor materials that contains all the functional components and connections of the circuit, integrated into a single device package.

ICC (integrated circuit card) Another name for a smart card.

ICMA (International Card Manufacturers Association Suite) A smart card industry trade association. For more information, go to www.icma.com.

IDEA A cryptographic algorithm commonly thought of as the European equivalent of DES.

IEC (International Electrotechnical Commission) An international standards body based in Geneva, Switzerland.

IFD (interface device) Another name for a smart card reader.

induced error attack An attack on a smart card's security system that causes the CPU to perform erroneous calculations; errors are induced in the smart card's CPU by subjecting the card to unusual environmental conditions such as temperature, voltage, microwaves, radiation, and so on.

initial bit The first bit of a string of bits presented to an input device. The device will group the series into blocks of, say, 8 bits to make a byte string. It is important to specify if the initial bit is the highest or lowest bit in its byte.

initialization The process during which the basic data that is common to all chip cards in a manufacturing batch is loaded into the chip.

INS The second field of an ISO 7816-4 smart card command, which contains the instruction to be executed by the smart card.

intelligent memory card A memory card that contains some additional features—typically security features—which limit access to the memory.

inverse convention A communication convention wherein signal-positive is to be interpreted as 0 and signal-zero is to be interpreted as 1; this is the inverse of the usual translation of these states into binary digits.

ISO (International Standards Organization) The penultimate technical standards body based in Geneva, Switzerland. With representation on its working committees from almost all countries, the ISO defines technical standards for worldwide interoperability of hardware and software. For more information, go to www.iso.org.

ISO/IEC 4909 The ISO standard for magnetic card format for electronic banking data. Some smart cards have magnetic strips on them and others support magnetic stripe communication protocols.

ISO/IEC 7810 The ISO standard for the physical characteristics of an identification card.

ISO/IEC 7811 The ISO standard for identification card recording techniques.

ISO/IEC 7812 The ISO standard encoding for identifying issuers of financial smart cards.

ISO/IEC 7813 The ISO standard that defines the specifics of financial transaction identification cards.

ISO/IEC 7816 The basic set of international standards covering smart cards. There are currently six parts to the ISO 7816 standard:

Part 1—Defines the physical characteristics of the card.

Part 2—Defines the dimensions and location of contacts on the card. It also prescribes the meaning of each contact.

Part 3—Defines the electronic signals and transmission protocols required as specified in Part 2.

Part 4—Defines the commands to read, write, and update data.

Part 5—Defines application identifiers (AIDs).

Part 6—Defines data encoding rules for applications.

ISO/IEC 8583 The ISO standard for financial transaction messages.

ISO/IEC 9992 The ISO standard that describes the method of communication between card and reader for financial transaction cards.

ISO/IEC 10181-3 The ISO standard for access control.

ISO/IEC 10202 The ISO standard for the architecture of the systems that utilize financial transaction cards.

ISO/IEC 10373 The ISO standard for testing smart cards.

ISO/IEC 10536 The basic ISO standard for contactless smart cards.

ISO/IEC JTC1/SC17 The ISO standing committee responsible for smart card standards. For more information, go to www.iso.ch/meme/JTC1SC17.html.

issuer The institution or organization that creates, provides, and typically owns a smart card.

Java Card A smart card that includes a Java interpreter in its operating system.

For more information, go to www.javasoft.com.

Java Card Forum An organization of smart card manufacturers that offer Java smart cards. For more information, go to www.javacardforum.org.

KLOC One thousand lines of code.

layout The organization of dedicated and elementary files in the smart card's EEPROM.

linear file A type of file in an ISO 7816-4 smart card file system that contains records. The records in a linear file may be fixed length or variable length.

loyalty program A product-marketing scheme that entices customers to purchase the product repeatedly by offering rewards based on the frequency of purchase. Also known as frequent buyer programs or, from its airline origin, frequent flyer programs.

MAC (message authentication code) A cryptographic checksum used to detect whether text of or data in the message has been modified.

MAOS (multiapplication operating system) A smart card operation system licensed by MAOSCO that is also known as MULTOS. For more information, go to www.multos.com.

mask The program written into a smart card chip's ROM during its manufacture; typically, the smart card's operating system and manufacturer's data.

memory card A plastic card with a simple memory chip with read and write capability.

memory chipcard A memory card in which access to the data in the EEPROM is controlled by security logic. *See also* intelligent memory card.

MF (master file) The root directory of a smart card's file system. An MF can contain dedicated files (other directories) and elementary files (data files). The master file on an ISO 7816–compliant smart card has the file identifier $3F00_{16}$.

MFC (multifunction card) A smart card that contains more than one application.

MIP Million instructions per second.

module The metal carrier into which a smart card chip is placed before it is embedded into a plastic body to make a smart card. The module provides mechanical protection for the chip and contains the contacts or pads that a smart card reader connects to in order to activate and communicate with the chip.

MONDEX A smart card operating system developed by NatWest in the UK and also an e-cash smart card that supports direct transfer of value from one card to another. For more information, go to www.mondex.com.

Montgomery multiplication An efficient way to do binary multiplication based on shifting and adding.

Montgomery multiplication is particularly useful in multiplying the arbitrarily large integers used in some cryptographic algorithms on the 8-bit microcontroller in a smart card.

Moréno, Roland The French journalist who received a patent on smart cards in 1974.

mother card A smart card holding a transport key and used to unlock all the cards in a batch or shipment of cards. *See also* batch card, daughter card.

MULTOS The multiapplication smart card operating system on the MONDEX card and licensable from MAOSCO to be the foundation for any multiapplication smart card. For more information, go to www.multos.com.

NACCU (National Association of Campus Card Users) A North American smart card industry group. For more information, go to www.naccu.org.

NIST (National Institute for Standards and Technology) An American standards body particularly for the use of information-processing technology by the Federal government. For more information, go to www.nist.gov.

NVM (nonvolatile memory) A generic term for the memory in a smart card that can be written but still holds its contents after power has been removed; PROM, EPROM, EEPROM, FLASH, and FRAM are examples of NVM.

offline The state in which a smart card is not connected to a computer network and must rely on the information stored in its own file system; for example, to approve or deny a transaction.

online The state in which a smart card is connected to a computer network and can be instructed to, for example, accept or deny a transaction based on information it sends to computers on the network.

optical card A memory card that can be written once but read many times and can hold between 1 MB and 40 MB of data. Reading and writing use laser optical technology.

page size The smallest number of bytes in EEPROM memory that can be written with one write operation. Page sizes in smart cards vary from 1 to 32 bytes.

path The location of a file with respect to the root directory.

PC/SC (personal computer/smart card) A group of personal computer and smart card companies, founded to work on open specifications to integrate smart cards with personal computers. For more information, go to www.smartcardsys.com.

personalization The process during which individual data are loaded into the smart card chip. Typically performed together with the printing or embossing of personal data (name, ID number, picture, and so on) and an account number onto the face of the card.

phone card A card that can be used for the payment of telephone charges, typically in a pay phone.

PIN (personal identification number) Typically a four- or five-digit number used by the operating system on the smart card to authenticate the cardholder.

PKA (public key algorithm) A cryptographic algorithm that uses a pair of keys, a public key and a private key, that are different from one another. The public key is published and available to anyone wishing to send an encrypted communication to the holder of the private key. *See also* SKA (secret key algorithm).

PKI (public key infrastructure) A system of storing and distributing public keys together with their current status, typically at scale (that is, millions to billions of keys).

POS (point of sale) A type of terminal found, for example, at grocery store checkout stations.

private key A cryptographic key known only to the owner. Or, the secret component of an asymmetric cryptographic key. *See also* PKA (public key algorithm).

processor card A smart card that contains a microprocessor or microcontroller that can execute a program stored in the card's memory.

processor core *See* core.

Proton A smart card operating system developed by Banksys in Belgium. Used for travel and entertainment by American Express, Hilton Hotels, and American Airlines in the United States and for e-cash in Sweden. For more information, go to www.proton.be.

public key The publicly available and distributed component of an asymmetric cryptographic key.

purse file A type of file in a smart card's file system that is used to implement electronic purses.

PVC (polyvinyl chloride) Plastic material used for the body of some smart cards.

RAM (random access memory) Memory used for temporary storage of data by the CPU in a smart card. RAM is volatile; its contents are lost when power is removed from the smart card. *See also* NVM (nonvolatile memory).

Regulation E A U.S. Federal regulation designed to protect users and issuers using electronic financial transfers from fraudulent transactions. It requires users to receive a receipt of financial transactions, puts restrictions on issuance of accessible devices, establishes the conditions of this type of service, and puts limits on consumer liability.

relative path The location of a file relative to the current file.

retention time The length of time a smart card will hold data in its non-volatile memory—typically, 10 years.

RF/DC (radio frequency/direct communication) A method of communication without physical contact using radio frequency transmission.

RF/ID (radio frequency/identification) A method identification without physical contact using radio frequency transmission.

ROM (read-only memory) A permanent memory in a smart card to which the CPU cannot write new information and that cannot be updated or changed. It is written during the manufacture of the chip and typically contains the smart card operating system and manufacturer keys.

RSA An asymmetric cryptographic algorithm named after its inventors, Rivest, Shamir, and Adleman. For more information, go to www.rsa.com.

RST The contact or pad on the smart card module that, when activated, causes a physical reset of the microprocessor in the smart card.

SDK (software development kit) A collection of software and software tools useful in building a particular kind of software application; (such as a smart card software development kit or a graphics software development kit).

SET (secure electronic transactions) A protocol developed by Visa and MasterCard for making credit card purchases on the Internet.

SIM (subscriber identity module) The type of module used in GSM smart cards to allow personal access to the GSM network. The SIM contains the user's cellular telephone account information.

simulator A computer program that runs on a personal computer, for example, that executes a program to eventually be executed on a smart card and provides tools to the smart card program developer to study and debug the smart card program. *See also* emulator.

SKA (secret key algorithm) A cryptographic algorithm that uses a single key that is shared by the sender and the recipient of the encrypted message. The single key is used for both encryption and decryption and must be kept a secret shared between them.

smart card A plastic card with a microprocessor chip that provides secure access to the memory of the card and performs other data-processing and communication functions. Smart cards are used to store monetary value and personal identification information.

smart card editor A program typically with a graphical user interface that enables you to see and change the contents of a smart card as well as send the smart card any command it supports.

Smart Card Forum A smart card trade association. See www.smartcrd.com.

soft mask Executable code typically written in machine language that is written into a smart card's nonvolatile memory after the card is manufactured. Soft mask code can either correct errors in the smart card operating system stored in ROM or add additional capabilities to the smart card.

SPOM (self-programmable) A one-chip microcomputer in which one integrated circuit contains all the electronic components of the microcomputer. Smart card chips are SPOMs.

stamp A MAC additionally containing input data.

start bit In an asynchronous communication protocol, the start bit signals the beginning of a new message and alerts the receiver to start collecting the bits of the message. The start bit typically serves only this heads-up function and is not part of the message itself.

SVC (stored value card) A smart card that stores nonbearer values such as e-cash. Some stored value cards can be reloaded with more value and some cannot.

swallow To pull the smart card completely inside the reader so that the cardholder can't remove the card from the reader during a transaction.

symmetric algorithms A cryptographic algorithm or protocol in which the same key is held by both parties and is used for both encryption and decryption. DES is a symmetric algorithm.

symmetric key A cryptographic key used in a symmetric cryptographic algorithm. It is called *symmetric* because the same key is used to decrypt a message as was used to encrypt the message. *See also* SKA (secret key algorithm).

synchronous protocol A communication protocol that is premised on the existence of a common clock or synchronized clocks between the sender and the receiver of the data.

T=0 A communication protocol between a smart card and a smart card reader thatt ransfers information one byte at a time; a byte-oriented smart card communication protocol.

T=1 A communication protocol between a smart card and a smart card reader that transfers information in a block or blocks of multiple bytes; a block-oriented smart card communication protocol.

tamper detection Capabilities of a smart card such as low voltage or slow clock detection circuits that enable the card to detect an attempted, unauthorized access to data it contains or an attempt to alter the calculations it performs.

tamper evident Physical aspects of a smart card that, when altered, will not return to their unaltered state and thus will show that the card has been tampered with.

tamper resistant Properties of a smart card—both in hardware and software—that make it difficult to perform unauthorized alterations of the data stored in the smart card or to make the smart card perform unauthorized computations.

tamper response Actions such as zeroization taken by a smart card when tampering is detected.

TE (terminal equipment) Another name for a smart card reader.

tear To remove a smart card from the smart card reader in the middle of a transaction; may leave the data on the smart card in an inconsistent or incorrect state.

TESA-7 A cryptographic algorithm used in GSM telephony.

timing attack An attack on a smart card's security system that is based on precise measurement of how long it takes the microprocessor to perform certain functions. For example, it takes longer to multiply by one than by zero.

TLV (tag length value) A way of formatting arbitrary data for transmission between a smart card and a host application.

TPDU (Transmission Protocol Data Unit) A block of data sent from the smart card to the host application.

transaction A business or payment event for the exchange of value for goods or services.

transaction time The amount of time between the start and finish of a transaction.

transparent file A type of file organization. The EEPROM file contains a byte string. Data is accessed using the offset length relative to the first byte within the byte string.

transportation key or transport key A key that prevents data being written into a smart card NVM when it is being transported from the chip manufacturer to the card manufacturer or from the card manufacturer to the card issuer.

Unicode A method for encoding characters from many alphabets in 2 bytes or 16 bits. For example, $03BE_{16}$ is the lowercase Greek letter epsilon ($\in$). *See also* ASCII.

value checker A battery-operated smart card reader for checking the current value held in a stored value card.

V_{CC} The contact or pad on a smart card module through which voltage is supplied to power the smart card processor; also the voltage itself, typically 5 volts.

Visa Cash card A stored-value smart card produced by Visa that transfers U.S. cash.

voltage attack An attack on a smart card's security system that is based on making very precise measurements of how much voltage the smart card draws. For example, some smart card chips draw more voltage when they are multiplying by one than when they are multiplying by zero.

V$_{PP}$ The contact or pad on a smart card module through which voltage is supplied to program or to erase the nonvolatile memory of the smart card; also, the voltage itself, typically 5 volts.

wired logic card *See* intelligent memory card.

write/erase time The amount of time it takes to write or erase a page of nonvolatile memory in a smart card. Typically on the order of 5 milliseconds for EEPROM memory.

zeroization Setting the nonvolatile memory of a smart card to all null values (zero), wiping out all data stored on the smart card; typically done in response to tamper detection.

Index

Symbols

A

functions

G

N

O

P

S

security

software

MACMILLAN COMPUTER PUBLISHING USA

A VIACOM COMPANY

Technical
Support

If you need assistance with the information
in this book or with a CD/disk accompanying
the book, please access the Knowledge Base
on our Web site at **http://www.superlibrary.
com/general/support**. Our most frequently
asked questions are answered there. If you
do not find the answer to your questions on
our Web site, you may phone Macmillan
Technical Support at **(317) 581-3833** or e-mail
us at **support@mcp.com**.

Now That You've Studied Smart Card Programming, It's Time to Get Started

MTP

MACMILLAN
TECHNICAL
PUBLISHING
U·S·A

Schlumberger offers smart card readers and development tools that you will need to continue to do your own card formatting and to develop your own smart card applications.

Schlumberger smart card developer products include

- **Cyberflex Development Kit,** for programming Java cards, based on Schlumberger's Cyberflex smart card, the first card for Java
- **SafePaK,** for developing PC applications based on Cryptoflex, the first smart card that brings hardware-strength RSA security to your applications
- **EZ Formatter,** a card formatting tool (included in the Cyberflex Development Kit)

- **EZ Component,** an add-on program for the Delphi Windows programming environment (included in the Cyberflex Development Kit)
- The **Reflex 20** smart card reader for PCMCIA slot
- The **Reflex 60** smart card reader for serial port connection, with or without built-in keypad
- The **Litronic Argus 210** smart card reader for serial port connection
- **Developer packs** of smart cards (packs of five each Multiflex, Cyberflex, or Cryptoflex cards)

You can order them at
`http://www.cyberflex.austin.et.slb.com`
or by calling 800-825-1155
or emailing `lambert@owings-mills.et.slb.com`.

Schlumberger